Study Guide for
SOCIOLOGY IN OUR TIMES

Diana Kendall
Austin Community College

Wadsworth Publishing Company
I(T)P® An International Thomson Publishing Company

Belmont • Albany • Bonn • Boston • Cincinnati • Detroit • London • Madrid • Melbourne
Mexico City • New York • Paris • San Francisco • Singapore • Tokyo • Toronto • Washington

Printed in the United States of America
1 2 3 4 5 6 7 8 9 10

For more information, contact Wadsworth Publishing Company.

Wadsworth Publishing Company
10 Davis Drive
Belmont, California 94002, USA

International Thomson Publishing Europe
Berkshire House 168-173
High Holborn
London, WC1V 7AA, England

Thomas Nelson Australia
102 Dodds Street
South Melbourne 3205
Victoria, Australia

Nelson Canada
1120 Birchmount Road
Scarborough, Ontario
Canada M1K 5G4

International Thomson Editores
Campos Eliseos 385, Piso 7
Col. Polanco
11560 México D.F. México

International Thomson Publishing GmbH
Königswinterer Strasse 418
53227 Bonn, Germany

International Thomson Publishing Asia
221 Henderson Road
#05-10 Henderson Building
Singapore 0315

International Thomson Publishing Japan
Hirakawacho Kyowa Building, 3F
2-2-1 Hirakawacho
Chiyoda-ku, Tokyo 102, Japan

ISBN 0-534-21027-9

TABLE OF CONTENTS

CHAPTER 16: POPULATION AND URBANIZATION

CHAPTER 17: COLLECTIVE BEHAVIOR AND SOCIAL CHANGE

PREFACE

Welcome to the study of *Sociology in Our Times*! This study guide has been written to assist you in mastering the material in your text and doing well on your instructor's tests. Each chapter in the Study Guide corresponds with the same chapter in *Sociology in Our Times* and has the following components:

Brief Chapter Outline

Chapter Summary

Learning Objectives

Key Terms

Key People

Chapter Outline

Analyzing and Understanding the Boxes

Practice Test

Sociology in Our Times: Diversity Issues

Crossword Puzzle

How To Use This Study Guide

1. Before reading each chapter, look over the Brief Chapter Outline and Chapter Summary to get an overview of the topics in each chapter.

2. As you read, use the Learning Objectives, Key Terms, and Key People as a guide for taking notes. (You may wish to start a sociology notebook with concise responses for study purposes.)

3. Look for recurring themes in each chapter and between chapters. For example, how do functionalist, conflict, and interactionist perspectives view each of the topics being examined? Outline the key components of each theory in your notebook.

4. Look at the topics and issues presented through the lenses of race, class, gender, age, and (where applicable) ability/disability.

5. In the section on "Analyzing and Understanding the Boxes," outline the key points and possible discussion questions for each of the four standard boxes -- Sociology in Everyday Life, Sociology and Media, Sociology and Law, and Sociology in Global Perspective --that are included in each chapter.

6. <u>After you have read</u> the chapter, take the practice test under actual testing situations (i.e., "closed book," sitting at a table, and timing yourself). Grade your practice test using the answers at the end of each Study Guide chapter. Use the page numbers indicated on the practice test and in the answers to revisit the text for your information on each item you missed. (You may wish to save the practice tests to use as a self-test before you take an in-class examination.)

7. Work the crossword puzzle for fun and to check your comprehension of key concepts, people, and theories.

To optimize your use of the *Study Guide to Accompany Sociology in Our Times*, follow the reading schedule set forth by your professor. Read each chapter and complete the activities in the study guide <u>before</u> you hear your instructor's lecture and/or discussion of the topics. If you are taking a distance-learning (self-paced) course, this study guide will be even more important to you in organizing the content in each chapter and making sure that you understand the key components.

Best wishes as you explore *Sociology in Our Times*. If you would like to comment on the text, this study guide, or any aspect of sociology, please write me (and I will respond) at this address:

Dr. Diana Kendall
Austin Community College
1212 Rio Grande
Austin, Texas 78701
e-mail address: dkendall@austin.cc.tx.us.edu

CHAPTER 1
THE SOCIOLOGICAL PERSPECTIVE

BRIEF CHAPTER OUTLINE
Putting Social Life into Perspective
>   Why Study Sociology?
>   The Sociological Imagination

The Development of Sociological Thinking
>   Early Thinkers: A Concern with Social Order and Stability
>   Differing Views on the Status Quo: Stability vs. Change
>   Development of Sociology in the United States

Contemporary Theoretical Perspectives
>   Functional Perspectives
>   Conflict Perspectives
>   Interactionist Perspectives

CHAPTER SUMMARY
Sociology is the systematic study of human society and social interaction. Sociology enables us to see how individual behavior is largely shaped by the groups to which we belong and the **society** in which we live.   The **sociological imagination** helps us to understand how seemingly personal troubles, such as rape victimization, actually are related to larger social forces. Sociology emerged out of the social upheaval produced by **industrialization** and **urbanization** in the late eighteenth century. Some early social thinkers -- including **Auguste Comte, Harriet Martineau, Herbert Spencer**, and **Emile Durkheim** -- emphasized social order and stability; others -- including **Karl Marx, Max Weber**, and **Georg Simmel** -- focused on conflict and social change. From its origins in Europe, sociology spread to the United States in the 1890s when departments of sociology were established at the University of Chicago and Atlanta University. Sociologists use three primary theoretical perspectives to examine social life: (1) **functionalist perspectives** assume that society is a stable, orderly system; (2) **conflict perspectives** assume that society is a continuous power struggle among competing groups, often based on **class, race, ethnicity**, or **gender**; and (3) **interactionist perspectives** focus on how people make sense of their everyday social interactions.

LEARNING OBJECTIVES
After reading Chapter 1, you should be able to:
1.   Describe the sociological imagination and explain its importance in understanding people's behavior.

2. Explain what C. Wright Mills meant by the sociological imagination and why it requires us to include many points of view and diverse experiences in our own thinking.

3. Define race, ethnicity, class, sex, and gender, and explain why these terms are important to the development of our sociological imaginations.

4. Discuss industrialization and urbanization as factors that contributed to the development of sociological thinking.

5. Identify Auguste Comte, Harriet Martineau, and Herbert Spencer, and explain their unique contributions to early sociology.

6. Contrast Emile Durkheim's and Karl Marx's perspectives on society and social conflict.

7. Explain what Max Weber meant by value free and Verstehen. What part do these terms play in the sociological imagination?

8. Describe the origins of sociology in the United States and discuss the role of women in early departments of sociology and social work.

9. State the major assumptions of functionalism, conflict theory, and interactionism, and identify the major contributors to each perspective.

10. Describe the views of Max Weber, Ralf Dahrendorf, and C. Wright Mills on power and who holds it.

11. Distinguish between microlevel and macrolevel analyses and state which level of analysis is utilized by each of the major theoretical perspectives.

12. State the key assumptions of the interactionist perspective.

KEY TERMS (defined at page number shown and in glossary)

alienation 19
anomie 18
bourgeoisie 19
class 12
class conflict 19
commonsense knowledge 6
conflict perspectives 26
dysfunctions 25
ethnicity 12
functionalist perspectives 24
gender 12
global interdependence 6
group consciousness 13
industrialization 14
interactionist perspectives 29
latent functions 25
macrolevel analysis 29
manifest functions 25
means of production 19

microlevel analysis 29
objective 9
perspective 23
positivism 16
power elite 27
proletariat 19
race 12
sex 12
social Darwinism 17
social disorganization 22
social facts 17
societal consensus 24
society 6
sociological imagination 9
sociology 5
symbol 29
theory 23
urbanization 14

KEY PEOPLE (identified at page number shown)

Jane Addams 22
Ernest Burgess 22
August Comte 15
W.E.B. Du Bois 22
Emile Durkheim 17
Harriet Martineau 16
Karl Marx 18
George Herbert Mead 22

C. Wright Mills 9
Robert E. Park 22
Talcott Parsons 25
Georg Simmel 20
Albion Small 21
Herbert Spencer 17
Max Weber 19

CHAPTER OUTLINE

I. PUTTING SOCIAL LIFE INTO PERSPECTIVE
   A.   **Sociology** is the systematic study of human society and social interaction.
   B.   Why Study Sociology?
       1.   Sociology helps us see the complex connections between our own lives and the larger, recurring patterns of the society and world in which we live.
          a.   A **society** is a large social grouping that shares the same geographical territory and is subject to the same political authority and dominant cultural expectations.
          b.   When we examine the world order, we become aware of **global interdependence** -- a relationship in which the lives of all people are intertwined closely and any one nation's problems are part of a larger global problem.
          c.   Sociological research often reveals the limitations of myths associated with **commonsense knowledge** that guides ordinary conduct in everyday life.
   C.   The Sociological Imagination
       1.   According to sociologist C. Wright Mills, the **sociological imagination** enables us to distinguish between personal troubles and public issues.
       2.   Developing a "personal" sociological imagination requires that we take into account perspectives of people from diverse backgrounds.
          a.   People in the United States differ by **race** -- a term used by many people to specify groups of people distinguished by physical characteristics such as skin color -- and **ethnicity** -- cultural heritage or identity of a group, based on factors such as language or country of origin.
          b.   They also differ by **class** -- the relative location of a person or group within a larger society, based on wealth, power, prestige, or other valued resources -- and by **gender** -- the meanings, beliefs, and practices associated with sex differences.

II. THE DEVELOPMENT OF SOCIOLOGICAL THINKING
   A.   **Industrialization** -- the process by which societies are transformed from dependence on agriculture and handmade products to an emphasis on manufacturing and related industries

4

-- and **urbanization** -- the process by which an increasing proportion of a population lives in cities rather than rural areas -- contributed to the development of sociological thinking.

B.   Some early social thinkers were concerned with social order and stability:

1.   **Auguste Comte** coined the term sociology and stressed the importance of **positivism** -- a belief that the world can best be understood through scientific inquiry.

2.   **Harriet Martineau's** most influential work was *Society in America* in which she paid special attention to U.S. diversity based on race, class, and gender.

3.   **Herbert Spencer** used an evolutionary perspective to explain stability and change in societies. He coined the term "survival of the fittest" and became known for **social Darwinism** -- the belief that those species of animals, including human beings, best adapted to their environment survive and prosper, while those poorly adapted die out.

4.   According to **Emile Durkheim**, **social facts** are patterned ways of acting, thinking, and feeling that exist outside any one individual and exert social control over each person. **Anomie** is a condition in which social control becomes ineffective as a result of the loss of shared values and of a sense of purpose in society.

C.   Other early theorists had differing views on the status quo and stability vs. change:

1.   **Karl Marx** believed that conflict -- especially class conflict -- is inevitable.

   a.   Class conflict is the struggle between members of the **capitalist class**, or **bourgeoisie** and the **working class**, or **proletariat**.

   b.   Exploitation of workers by capitalists results in workers' **alienation** -- a feeling of powerlessness and estrangement from other people and from oneself.

2.   In *The Protestant Ethic and the Spirit of Capitalism*, **Max Weber** suggested that religion is a central force in social change.

3.   **Georg Simmel** emphasized that society is best seen as a web of patterned interactions that make up the "geometry of social life."

D.   The first U.S. department of sociology was at the University of Chicago.

    1.   **Social disorganization** refers to conditions that undermine the ability of traditional institutions (such as family, church, or school) to govern social behavior.

    2.   **Jane Addams** wrote *Hull-House Maps and Papers* which was used by other Chicago sociologists for the next forty years.

    3.   **W.E.B. Du Bois** founded the second U.S. department of sociology at Atlanta University and wrote *The Philadelphia Negro: A Social Study*, examining Philadelphia's African American community.

III. CONTEMPORARY THEORETICAL PERSPECTIVES

A.   A **theory** is a set of logically interrelated statements that attempts to describe, explain, and (occasionally) predict social events. Theories provide a framework or **perspective** -- an overall approach or viewpoint toward some subject -- for examining various aspects of social life.

B.   **Functionalist perspectives** are based on the assumption that society is a stable, orderly system characterized by societal consensus.

    1.   Societies develop social structures, or institutions, that persist because they play a part in helping society survive. These institutions include: the family, education, government, religion, and economy.

    2.   **Talcott Parsons** stressed that all societies must make provisions for meeting social needs in order to survive. For example, a division of labor (distinct, specialized functions) between husband and wife is essential for family stability and social order.

    3.   Robert K. Merton distinguished between intended and unintended functions of social institutions.

        a.   **Manifest functions** are intended and/or overtly recognized by the participants in a social unit.

        b.   **Latent functions** are unintended functions that are hidden and remain unacknowledged by participants.

        c.   **Dysfunctions** are the undesirable consequences of any element of society.

C.   According to **conflict perspectives**, groups in society are engaged in a continuous power struggle for control of scarce resources.

    1.   Along with Karl Marx, Max Weber believed that economic conditions were important in producing inequality and conflict in society; however, Weber also

suggested that power and prestige are other sources of inequality.

    2.   **Ralf Dahrendorf** observed that conflict is inherent in all authority relationships.

    3.   **C. Wright Mills** believed that the most important decisions in the United States are made largely behind the scenes by the **power elite,** a small clique composed of the top corporate, political, and military officials.

    4.   Feminist perspectives focus on patriarchy -- a system in which men dominate women, and that which is considered masculine is more highly valued than that which is considered feminine.

D.   Functionalist and conflict perspectives focus primarily on **macrolevel analysis** -- an examination of whole societies, large-scale social structures, and social systems. By contrast, interactionist approaches are based on a **microlevel analysis** -- an examination of everyday interactions in small groups rather than large-scale social structures.

E.   **Interactionist perspectives** are based on the assumption that society is the sum of the interactions of individuals and groups.

    1.   **George Herbert Mead**, a founder of this perspective, emphasized that a key feature distinguishing humans from other animals is the ability to communicate in **symbols** -- anything that meaningfully represents something else.

    2.   Some interactionists focus on people's behavior while others focus on each person's interpretation or definition of a given situation.

F.   All three theoretical perspectives contribute to our understanding of human behavior in social groups.

ANALYZING AND UNDERSTANDING THE BOXES

After reading the chapter and studying the outline, re-read the four boxes and write down key points and possible questions for class discussion.

Sociology and Law -- "The Issue of Consent"

Key Points:

Discussion Questions:

1.

2.

3.

Sociology and Everyday Life -- "How Much Do You Know About Rape?"

Key Points:

Discussion Questions:

1.

2.

3.

Sociology in Global Perspective -- "Rape in Wartime"

Key Points:

Discussion Questions:

1.

2.

3.

Sociology and Mass Media -- "Rape in Prime-Time"

Key Points:

Discussion Questions:

1.

2.

3.

PRACTICE TEST

MULTIPLE CHOICE QUESTIONS
Select the response that best answers the question or completes the statement:

1. Sociology is the systematic study of: (p. 5)
    a.   intuition and commonsense knowledge.
    b.   human society and social interaction.
    c.   the production, distribution, and consumption of goods and services in a society.
    d.   personality and human development.

2. All of the following are reasons to study sociology, except: (p. 6)
    a.   Sociology helps us gain a better understanding of ourselves and our social world.
    b.   Sociology is based on the use of scientific standards to study society.
    c.   Sociology confirms the accuracy of media depictions of social life.
    d.   Sociology helps us look beyond our personal experiences and gain insights into society.

3. The text points out that ordinary conduct in everyday life is frequently guided by: (p. 6)
    a.   sociological concepts.
    b.   psychiatric principles.
    c.   media commentary.
    d.   commonsense knowledge.

4. According to C. Wright Mills, the sociological imagination refers to the ability to: (p. 9)
    a.   distinguish between personal troubles and public issues.
    b.   see the relationship between preliterate and literate societies.
    c.   be completely objective in examining social life.
    d.   seek out one specific cause for a social problem such as rape.

5. All of the following statements reflect a "rape-supportive" belief system in the United States, except: (p. 10)
    a.   Men should be more aggressive and physically assertive than women.
    b.   Women are "asking for it" when they wear a short skirt or go out alone at night.
    c.   Date rape happens when the victim does not say "no" loud enough.
    d.   Rape is a product of society, not merely a personal problem.

6. Femininity and masculinity are _____ - related terms. (p. 12)
   a. sex
   b. gender
   c. biologically
   d. sociobiologically

7. According to sociologists, an awareness that an individual's problems are shared by others who are similarly situated in regard to race, ethnicity, gender, class, or age is: (p. 13)
   a. personal awareness.
   b. social awareness.
   c. individual prejudice.
   d. group consciousness.

8. Two historical factors that contributed to the development of sociological thinking were: (p. 14)
   a. industrialization and urbanization.
   b. industrialization and immigration.
   c. urbanization and centralization.
   d. urbanization and immigration.

9. Each of the following people made an important contribution to early sociology, except: (p. 15)
   a. Auguste Comte.
   b. Harriet Martineau.
   c. Adam Smith.
   d. Herbert Spencer.

10. _____ argued that conflict -- especially class conflict -- is necessary in order to produce social change and a better society (p. 18).
   a. Auguste Comte
   b. Emile Durkheim
   c. Karl Marx
   d. Harriet Martineau

11. According to Max Weber, sociology should be "value free," meaning that: (p. 20)
   a. sociologists should have no values of their own.
   b. sociological research should exclude the researcher's own personal values and economic interests.
   c. sociologists should ignore the values of people who participate in their research.
   d. sociologists should not employ *verstehen* to gain the ability to see the world as others see it.

12. The first U.S. department of sociology was established at _____, where _____ was one of the best known women in the field. (p. 21-22).
    a.   University of Chicago - Jane Addams
    b.   Harvard University - Harriet Martineau
    c.   University of California - Arlie Hochschild
    d.   Princeton University - Sara McLanahan

13. W. E. B. Du Bois referred to the identity conflict of being a black and an American as: (p. 23)
    a.   group consciousness.
    b.   the American dilemma.
    c.   false consciousness.
    d.   double-consciousness.

14. _____ perspectives are based on the assumption that society is a stable, orderly system. (p. 24)
    a.   Functionalist
    b.   Interactionist
    c.   Conflict
    d.   Feminist

15. _____ perspectives are based on the assumption that groups are engaged in a continuous power struggle for control of scarce resources. (p. 26)
    a.   Functionalist
    b.   Interactionist
    c.   Conflict
    d.   Feminist

16. According to your text, all of the following are conflict theorists, except: (p. 26)
    a.   Max Weber.
    b.   Talcott Parsons.
    c.   Ralf Dahrendorf.
    d.   C. Wright Mills.

17. The _____ approach directs attention to women's experiences and the importance of gender as an element of social structure. (p. 27).
    a.   functionalist
    b.   interactionist
    c.   conflict
    d.   feminist

18. The relationship between social learning and rape is examined by sociologists using the _____ perspective. (p. 29)
    a.   macrolevel
    b.   functionalist
    c.   conflict
    d.   interactionist

19. According to the _____ perspective, society is the sum of the interactions of individuals and groups. (p. 29)
    a.   functionalist
    b.   interactionist
    c.   conflict
    d.   feminist

20. Signs, gestures, written language, and shared values are all examples of: (p. 29)
    a.   symbols.
    b.   psychological defense mechanisms.
    c.   norms.
    d.   roles.

TRUE-FALSE QUESTIONS

T F   1.   In evaluating the question, "Why should we study sociology?" the text points out that sociologists strive to use scientific standards in studying society and social interaction. (p. 9)

T F   2.   Today, the "pop" sociology of the mass media is quite similar to the discipline of sociology. (p. 9)

T F   3.   As the rate of reported rape increases, more people are becoming aware that rape is a public issue, not merely a personal problem. (p. 10)

T F   4.   According to sociologists, "pure" racial types are composed of people with distinctive physical characteristics such as skin color. (p. 12)

T F   5.   Auguste Comte is considered by some to be the "founder of sociology." (p. 15)

T F   6.   Herbert Spencer's view that societies developed through a process of "struggle" and "fitness" is known as "social Darwinism." (p. 17)

T   F   7.   Karl Marx developed the theory of anomie to explain how strains in society lead to a breakdown in traditional organizations and values. (p. 18)

T   F   8.   In the Marxian framework, class conflict is the struggle between the capitalist class and the working class. (p. 19)

T   F   9.   Max Weber's wife, Marianne Weber, was a radical feminist who lead a Marxist rebellion in Berlin. (p. 20)

T   F   10.   The first U.S. departments of sociology were located at the University of Chicago and at Atlanta University. (p. 21)

T   F   11.   Sociologically speaking, the terms "theory" and "perspective" mean the same thing. (p. 23)

T   F   12.   From a functionalist perspective, leadership, decision making, and employment outside the home to support the family, all constitute coordinating tasks. (p. 25)

T   F   13.   According to Robert K. Merton, manifest functions are intended and/or overtly recognized by the participants in a social unit. (p. 25)

T   F   14.   Conflict theorists who focus on racial-ethnic inequalities emphasize that rape cannot be studied without reference to the racism in which the history of rape is embedded. (p. 28)

T   F   15.   Interactionist perspectives are based on a macrolevel analysis of society. (p. 29)

SOCIOLOGY IN OUR TIMES: DIVERSITY ISSUES

1.   If both men and woman may be victimized by rape, why do you think we are more aware of women as victims of rape?

2.   How much do you rely on commonsense knowledge as a guide for your daily life? What insights do you think you may gain by studying sociology?

3.   What kinds of changes will occur in the United States over the next fifty years as the population grows by about 44 percent and becomes increasingly diverse? What impact will these changes have on your life?

4. How might a better understanding of the experiences of people who are quite dissimilar from you in regard to race/ethnicity, class, gender, and age help you in your personal life in the future? In your career?

5. The text states, that in our sociological imaginations, "we must include the experiences of people of all nations: those that are similar to the United States and those that are not." (p. 12) Why does the author think this is necessary? Do you agree? Why or why not?

6. Why were the sociological contributions of Harriet Martineau, Georg Simmel, Jane Addams, and W. E. B. Du Bois largely overlooked until recently? Do you think the contributions of some contemporary scholars may be overlooked or minimized because of their gender, race/ethnicity, religion, sexual orientation, age, disability, or other factors?

CHAPTER ONE CROSSWORD PUZZLE

For those who enjoy crossword puzzles, here is a puzzle that contains words and names from Chapter One. Working the puzzle will help you in reviewing the chapter. The answers appear on page 18.

ACROSS

1. The subject of this course
3. Ernest _____ was an early member of the "Chicago School," which was a group of sociologists at the University of Chicago from the 1920s to the 1940s
5. Jane _____, founder of Hull House
8. Karl Marx distinguished people and classes with regard to the _____
11. Sociology is the systematic study of _____ human society and social interaction
12. What may _____ to be a personal problem in fact may be a social issue
13. Sociology is the systematic study of human _____ and social interaction
14. He used an evolutionary perspective to explain social order and social change
16. Person who usually is thought of as being feminine
17. Gender refers to the meanings, beliefs, and practices associated with _____ differences
18. The relative location of a person or group within a larger society
19. Perspective which views groups in society as being engaged in a continuous power struggle for control of scarce resources
22. Functions that are intended and/or overtly recognized by the participants in a social unit
23. Industrialization and urbanization both represent _____ from an earlier situation or condition
24. Person considered by some to be the founder of sociology
26. The meanings, beliefs, and practices associated with femininity and masculinity
28. According to interactionist perspectives, society is the _____ of the interactions of individuals and groups
31. Social facts: patterned ways of acting, thinking, and feeling that exist outside _____ one individual
33. Opposite of functional

DOWN

1. Anything that meaning fully represents something else
2. To Marx, _____ conflict is the struggle between the bourgeoisie and the proletariat
3. Persons in the capitalist class
4. Common _____: what many people rely on to help understand our daily lives and other people's behavior
5. A condition in which social control becomes ineffective as a result of the loss of shared values and a sense of purpose in society
6. Founder of the second department of sociology in the United States
7. Popular but false notion that may be used, intentionally or unintentionally, to perpetuate certain theories
9. Patterned ways of acting, thinking and feeling that exist outside any one individual
10. First name of sociologist who analyzed the U.S. family in the 1950s on a functional basis
14. Founder of the first department of sociology in the U.S.
15. Term used by many people to specify groups of people distinguished by physical characteristics such as skin color
18. To Marx, this is inevitable when the capitalist class controls and exploits the masses
20. A set of logically interrelated statements that attempts to describe, explain, and (occasionally) predict social events
21. Sociologists _____ why and under what circumstances behavior takes place
22. Person who coined the term "the sociological imagination"
25. The person who emphasized that the ability to communicate in symbols is the key feature distinguishing humans from other animals
27. Conflict, functionalist, and interactionist persp_____ves
29. Adult male
32. A survey of high school students found that many male and female students believed that "a woman is willing to continue sexual activities even though she says _____"

ANSWERS TO PRACTICE TEST, CHAPTER 1

Answers to Multiple Choice Questions

1. b Sociology is the systematic study of human society and social interaction. (p. 5)
2. c All of the following are reasons to study sociology, <u>except</u>: Sociology confirms the accuracy of media depictions of social life. (p. 6)
3. d The text points out that ordinary conduct in everyday life is frequently guided by commonsense knowledge. (p. 6)
4. a According to C. Wright Mills, the sociological imagination refers to the ability to distinguish between personal troubles and public issues. (p. 9)
5. d All of the following statements reflect a "rape-supportive" belief system in the United States, <u>except</u>: Rape is a product of society, not merely a personal problem. (p. 10)
6. b Femininity and masculinity are gender - related terms. (p. 12)
7. d According to sociologists, an awareness that an individual's problems are shared by others who are similarly situated in regard to race, ethnicity, gender, class, or age is group consciousness. (p. 13)
8. a Two historical factors that contributed to the development of sociological thinking were industrialization and urbanization. (p. 14)
9. c All of the following people made an important contribution to early sociology, <u>except</u> Adam Smith. (p. 15)
10. c Karl Marx argued that conflict -- especially class conflict -- is necessary in order to produce social change and a better society. (p. 18)
11. b According to Max Weber, sociology should be "value free," meaning that sociological research should exclude the researcher's own personal values and economic interests. (p. 20)
12. a The first U.S. department of sociology was established at the University of Chicago, where Jane Addams was one of the best known women in the field. (p. 21-22).
13. d W. E. B. Du Bois referred to the identity conflict of being a black and an American as double-consciousness. (p. 23)
14. a Functionalist perspectives are based on the assumption that society is a stable, orderly system. (p. 24)
15. c Conflict perspectives are based on the assumption that groups are engaged in a continuous power struggle for control of scarce resources. (p. 26)
16. b According to your text, all of the following are conflict theorists, <u>except</u> Talcott Parsons. (p. 26)

16

17. d   The feminist approach directs attention to women's experiences and the importance of gender as an element of social structure. (p. 27).

18. d   The relationship between social learning and rape is examined by sociologists using the interactionist perspective. (p. 29)

19. b   According to the interactionist perspective, society is the sum of the interactions of individuals and groups. (p. 29)

20. a   Signs, gestures, written language, and shared values are all examples of symbols. (p. 29)

Answers to True-False Questions

1. True (p. 9)
2. False -- The mass media tend to look at singular events in isolation -- as individual and often bizarre occurrences; sociologists attempt to discover patterns or commonalities in human behavior. (p. 9)
3. True (p. 10)
4. False -- Sociologists argue that there is no such thing as "pure" racial types, and the concept of race is a myth. (p. 12)
5. True (p. 15)
6. True (p. 17)
7. False -- Emile Durkheim developed the theory of anomie to explain how strains in society lead to a breakdown in traditional organizations and values. (p. 18)
8. True (p. 19)
9. False -- Marianne Weber was an important figure in the women's movement in Germany during the early twentieth century; she also made her husband aware of many of the women's issues of his day. She did not lead a Marxist rebellion. (p. 20)
10. True (p. 21)
11. False -- A "theory" is a set of logically interrelated statements that attempts to describe, explain, and (occasionally) predict social events. Theories provide a framework in which observations may be logically ordered. Sociologists refer to their theoretical framework as a "perspective" -- an overall approach to or viewpoint on some subject. (p. 23)
12. False -- From a functionalist perspective, leadership, decision making, and employment outside the home to support the family, all constitute instrumental tasks. (p. 25)
13. True (p. 25)
14. True (p. 28)
15. False -- Interactionist perspectives are based on a microlevel analysis, which focuses on small groups rather than large-scale social structures. (p. 29)

```
S O C I O L O G Y     B U R G E S S   A D D A M S
Y   L               O         E     N   U   Y
M E A N S O F P R O D U C T I O N       O   B O T H
B   S O             R   A     S E E M   O     H
O   S O C I E T Y       G   L   E       I     I
L     I         S P E N C E R   S H E   S E X
      C L A S S     M   O     A
      O   L         A   I     T C O N F L I C T     A
M A N I F E S T     L   S     E               H     N
I   F   A           L   E         C H A N G E       A
L   L   C O M T E   G E N D E R             O       L
L   I   T   E               C         S U M R       I
S   C   S   A           I           T       A N Y   Z
    T       D Y S F U N C T I O N A L     N O     H E
```

CHAPTER 2
SOCIOLOGICAL RESEARCH METHODS

BRIEF CHAPTER OUTLINE
Why Is Sociological Research Necessary?
    Common Sense and Sociological Research
    Sociology and Scientific Evidence
    The Theory and Research Cycle
The Sociological Research Process
    Selecting and Defining the Research Problem
    Reviewing Previous Research
    Formulating the Hypothesis (If Applicable)
    Developing the Research Design
    Collecting and Analyzing the Data
    Drawing Conclusions and Reporting the Findings
    A Qualitative Approach to Researching Suicide
Research Methods for Collecting Data
    Experiments
    Surveys
    Secondary Analysis of Existing Data
    Field Research
    Multiple Methods of Social Research
Ethical Issues in Sociological Research
    The ASA Code of Ethics
    The Zellner Research
    The Humphreys Research
    The Scarce Research

CHAPTER SUMMARY
Social research is part of the sociological imagination. Sociologists conduct research to gain a more accurate understanding of society and provide a factual and objective counterpoint to commonsense knowledge and ill-informed sources of information. Sociological research is based on an **empirical approach** that answers questions through a direct, systematic collection and analysis of data. Theory and research form a continuous cycle that encompasses both **deductive** and **inductive approaches**. Many sociologists engage in **quantitative research**, which focuses on data that can be measured numerically. A researcher taking the quantitative approach might (1) select and define the research problem, (2) review previous research, (3) formulate the **hypothesis**, (4) develop the research design, (5) collect and analyze the data, and (6) draw conclusions and report the findings. Other research is qualitative, based on interpretive description rather than

statistics. Researchers taking the **qualitative approach** might (1) formulate the problem to be studied instead of creating a hypothesis, (2) collect and analyze the data and (3) report the results. **Research methods** -- systematic techniques for conducting research -- include **experiments, surveys, analyses of existing data, participant observation, complete observation, case studies, unstructured interviews,** and **ethnography.** Many sociologists use multiple methods in order to gain a wider scope of data and points of view. Studying human behavior raises important ethical issues for sociologists.

LEARNING OBJECTIVES
After reading Chapter 2, you should be able to:

1. Compare normative and empirical approaches to examining social issues.

2. Describe descriptive and explanatory studies and indicate how researchers decide which one to use.

3. Differentiate between quantitative and qualitative research and give examples of each.

4. Describe the six steps in the conventional research process, which focuses on deduction and quantitative research.

5. Indicate the relationship between dependent and independent variables in a hypothesis.

6. Distinguish between a representative sample and a random sample and explain why sampling is an integral part of quantitative research.

7. Explain why validity and reliability are important considerations in sociological research.

8. Describe the key steps in conducting qualitative research.

9. Contrast experimental and control groups and explain why control groups are necessary in experiments.

10. Describe the major types of surveys and indicate their major strengths and weaknesses.

11. State the major strengths and weaknesses of secondary analysis of existing data.

12. Describe the major methods of field research and indicate when researchers are most likely to utilize each of them.

13. Describe the major ethical concerns in sociological research.

KEY TERMS (defined at page number shown and in glossary)

complete observation 67
content analysis 66
control group 59
correlation 60
deductive approach 45
dependent variable 51
descriptive studies 44
empirical approach 43
ethnography 69
experiment 59
experimental group 59
explanatory studies 44
Hawthorne effect 62
hypothesis 51
independent variable 51
inductive approach 46
interview 63
normative approach 43

operational definition 53
participant observation 68
population 55
qualitative research 47
quantitative research 47
questionnaire 63
random sample 56
reliability 56
replication 58
representative sample 55
research methods 59
respondents 62
sample 55
secondary analysis 65
survey 62
unstructured interview 70
validity 56
variable 51

KEY PEOPLE (identified at page number shown)

Silvia Canetto 58
Emile Durkheim 38
Kevin E. Early 63
Myra Ferree and Elaine Hall 66
Barney Glaser and Anselm
    Strauss 71
Laud Humphreys 74

David Karp and William Yoels 67
Elliot Liebow 68
Robert and Helen Lynd 69
Rik Scarce 75
David Snow and Leon Anderson
    73
William Zellner 74

CHAPTER OUTLINE

I. WHY IS SOCIOLOGICAL RESEARCH NECESSARY?
    A. Sociologists obtain their knowledge of human behavior through research, which results in a body of information that helps us move beyond guesswork and common sense in understanding society.
    B. Sociology and Scientific Evidence
        1. Questions often are answered using one of two approaches:
            a. The **normative approach** uses religion, custom, habit, tradition, or law to answer important questions and focuses on what ought to be in society.
            b. The **empirical approach** attempts to answer questions through a systematic collection and

22

analysis of data, and thus is referred to as "the scientific method."

    2.   Empirical studies may be either descriptive or explanatory.

        a.   **Descriptive studies** attempt to describe social reality or provide facts about some group, practice, or event.

        b.   **Explanatory studies** attempt to explain cause and effect relationships and to provide information on why certain events do or do not occur.

  C.   The Theory and Research Cycle

    1.   A theory is a set of logically interrelated statements that attempts to describe, explain, and (occasionally) predict social events.

    2.   Research is the process of systematically collecting information for the purposes of testing an existing theory or generating a new one.

    3.   The theory and research cycle consists of the **deductive approach** and the **inductive approach**.

II. THE SOCIOLOGICAL RESEARCH PROCESS

  A.   Research may be either quantitative or qualitative.

    1.   **Quantitative research** is based on the goal of scientific objectivity and focuses on data that can be measured in numbers.

    2.   **Qualitative research** uses interpretive description (words) rather than statistics (numbers) to analyze underlying meanings and patterns of social relationships.

  B.   Steps in the research process include:

    1.   Selecting and defining the research problem;

    2.   Reviewing previous research;

    3.   Formulating the hypothesis (if applicable);

    4.   Developing the research design;

    5.   Collecting and analyzing the data; and

    6.   Drawing conclusions and reporting the findings.

  C.   Important concepts in the research process:

    1.   A **hypothesis** is a statement of the relationship between two or more concepts.

    2.   **Variables** are concepts with measurable traits or characteristics that can change or vary from one person, time, situation, or society to another.

    3.   The **independent variable** is presumed to cause or determine a dependent variable.

    4.   The **dependent variable** is assumed to depend on or be caused by the independent variable(s).

5. To use a variable, sociologists create an **operational definition** -- an explanation of an abstract concept in terms of observable features that are specific enough to measure the variable.

D. Important concepts in collecting and analyzing data:
1. The **population** consists of those persons about whom we want to be able to draw conclusions.
2. A **sample** is the people who are selected from the population to be studied, and should accurately represent that population.
    a. A **representative sample** is a selection from a larger population that has the essential characteristics of the total population.
    b. A **random sample** is chosen by chance: every member of an entire population being studied has the same chance of being selected.

E. Qualitative research differs from quantitative research in several ways:
1. Researchers do not always do an extensive literature search before beginning their investigation.
2. They may engage in problem formulation instead of creating a hypothesis.
3. This type of research often is built on a collaborative approach in which the "subjects" are active participants in the design process, not just passive objects to be studied.
4. Researchers tend to gather data in natural settings, such as where the person lives or works, rather than in a laboratory or other research setting.
5. Data collection and analysis frequently occur concurrently, and the analysis draws heavily on the language of the persons studied, not the researcher.

III. RESEARCH METHODS FOR COLLECTING DATA
A. **Research methods** are strategies or techniques for systematically conducting research.
B. **Experiments** -- carefully designed situations in which the researcher studies the impact of certain variables on subjects' attitudes or behavior -- typically require that subjects be divided into two groups:
1. The **experimental group** contains the subjects who are exposed to an independent variable (the experimental condition) to study its effect on them.
2. The **control group** contains the subjects who are not exposed to the independent variable.
3. The experimental and control groups then are compared to see if they differ in relation to the dependent variable,

and the hypothesis about the relationship of the two variables is confirmed or rejected.

    4.   Researchers acknowledge that experiments may have a problem known as the **Hawthorne effect** -- changes in the subject's behavior caused by the researcher's presence or by the subject's awareness of being studied.

C.   **Surveys** are polls in which researchers gather facts or attempt to determine the relationship between facts. Survey data are collected by using self-administered questionnaires, personal interviews, and/or telephone surveys.

    1.   A **questionnaire** is a printed research instrument containing a series of items for the subjects' response. Questionnaires may be self-administered by respondents or administered by interviewers in face-to-face encounters or by telephone.

    2.   An **interview** is a data-collection encounter in which an interviewer asks the respondent questions and records the answers. Survey research often uses structured interviews, in which the interviewer asks questions from a standardized questionnaire.

D.   In **secondary analysis of data**, researchers use existing material and analyze data that originally was collected by others.

    1.   Existing data sources include public records, official reports of organizations or government agencies, surveys taken by researchers in universities and private corporations, books, magazines, newspapers, radio, television, and personal documents.

    2.   **Content analysis** is the systematic examination of cultural artifacts or various forms of communication to extract thematic data and draw conclusions about social life.

E.   **Field research** is the study of social life in its natural setting: observing and interviewing people where they live, work, and play.

    1.   In **complete observation** researchers systematically observe a social process but do not become a part of it.

    2.   In **participant observation**, researchers collect systematic observations while being part of the activities of the groups they are studying.

    3.   A **case study** is an in-depth, multifaceted investigation of a single event, person, or social grouping.

    4.   **Ethnography** is a detailed study of the life and activities of a group of people by researchers who may live with that group over a period of years.

5.   An **unstructured interview** is an extended, open-ended interaction between an interviewer and an interviewee.
F.   Many sociologists utilize triangulation -- the use of multiple approaches in a single study.

IV. ETHICAL ISSUES IN SOCIOLOGICAL RESEARCH
A.   The study of people ("human subjects") raises vital questions about ethical concerns in sociological research.
B.   The American Sociological Association (ASA) has a Code of Ethics that sets forth certain basic standards sociologists must follow in conducting research.
C.   Sociologists are committed to adhering to this code and to protecting research participants; however, many ethical issues arise that cannot be resolved easily.

ANALYZING AND UNDERSTANDING THE BOXES
After reading the chapter and studying the outline, re-read the four boxes and write down key points and possible questions for class discussion.

Sociology and Everyday Life -- "How Much Do You Know About Suicide?"

Key Points:

Discussion Questions:

1.

2.

3.

Sociology and Media -- "Televising a Mass Suicide"

Key Points:

Discussion Questions:

1.

2.

3.

Sociology and Law -- "Assisting Suicide"

Key Points:

Discussion Questions:

1.

2.

3.

Sociology in Global Perspective -- "A Look at International Trends in Suicide"

Key Points:

Discussion Questions:

1.

2.

3.

PRACTICE TEST

MULTIPLE CHOICE QUESTIONS

Select the response that best answers the question or completes the statement:

1. According to Emile Durkheim, a high suicide rate is symptomatic of: (p. 38)
    a. individual problems.
    b. psychological abnormalities.
    c. small-scale group problems.
    d. a lack of cohesiveness in society.

2. The _____ approach uses religion, customs, habits, traditions, and law to answer important questions. (p. 43)
    a. normative
    b. empirical
    c. descriptive
    d. explanatory

3. Explanatory studies attempt to: (p. 44)
    a. use religion, customs, habits, traditions, and law to answer important questions.
    b. explain cause-and-effect relationships and to provide information on why certain events do or do not occur.
    c. describe social reality or provide facts about some group, practice, or event.
    d. interpret social events in an easy-to-understand manner.

4. A study of suicidal behavior based on suicide notes is an example of _____ research. (p. 47)
    a. quantitative
    b. qualitative
    c. non-scholarly
    d. triangulated

5. The first step in the research process is to: (p. 47)
    a. select and define the research problem.
    b. review previous research.
    c. develop a research design.
    d. formulate the hypothesis.

6. In Emile Durkheim's study of suicide, the degree of social integration was the: (p. 53)
    a.   operational definition.
    b.   dependent variable.
    c.   independent variable.
    d.   spurious correlation.

7. Suppose we are investigating the primary causes of suicide in the late 1990s; upon looking into recent cases of suicide, we find out that a number of the people had just lost their jobs; that they had been unemployed off and on for the past ten years; that they had no religious affiliation; and that a number of them had been divorced within the past five years. This analysis reflects what the text terms: (p. 54)
    a.   singular determination.
    b.   multivariate involvement.
    c.   plural association.
    d.   multiple causation.

8. A _____ sample refers to a situation in which every member of an entire population has the same chance of being selected for a study. (p. 56)
    a.   selective
    b.   random
    c.   representative
    d.   longitudinal

9. _____ is the extent to which a study or research instrument accurately measures what it is supposed to measure; _____ is the extent to which a study or research instrument yields consistent results. (p. 56)
    a.   Validity - replication
    b.   Replication - validity
    c.   Validity - reliability
    d.   Reliability - validity

10. Researchers following a qualitative approach to research: (p. 58)
    a.   always do an extensive literature search before beginning their investigation.
    b.   emphasize the importance of creating a hypothesis.
    c.   view the researcher as a "data-collecting machine."
    d.   attempt to formulate questions of concern and interest to the "subjects."

11. In an experiment, the subjects in the control group: (p. 59)
    a.   are exposed to the independent variable.
    b.   are not exposed to the independent variable.
    c.   are exposed to the dependent variable.
    d.   are not exposed to the dependent variable.

12. Suppose that the association of two variables is actually caused by a third variable, but a researcher assumes that the first variable causes the second. This illustrates a/an: (p. 61)
    a.   spurious correlation.
    b.   serendipity association.
    c.   by-chance relationship.
    d.   arbitrary association.

13. An important strength of self-administered questionnaires is: (p. 63)
    a.   integral complexity.
    b.   slow, methodical data collection.
    c.   low cost.
    d.   the high response rate.

14. Researchers who use existing material and analyze data that originally was collected by others are engaged in: (p. 65)
    a.   unethical conduct.
    b.   primary analysis.
    c.   secondary analysis.
    d.   survey analysis.

15. In the course of a research investigation designed to determine the popularity of exhibits at an art museum, we observe that there is more wear and tear on the floor in front of some exhibits. A content analyst would refer to these patterns as: (p. 66)
    a.   visual texts.
    b.   behavioral residues.
    c.   material culture.
    d.   nonmaterial culture.

16. Observation, ethnography, and case studies are examples of: (p. 67)
    a.   survey research.
    b.   experiments.
    c.   secondary analysis of existing data.
    d.   field research.

17. Sociologist Steve Taylor designed a study of coroners (medical examiners) to learn more about their rulings of "suicide" and to analyze what, if any, effect these rulings have on the accuracy of "official" suicide

statistics. In the process of his investigation, Taylor followed a number of cases from beginning to conclusion, listening to discussions and asking the coroners questions. Taylor's research may best be categorized as: (p. 68)
a.   participant observation.
b.   total observation.
c.   qualitative research.
d.   case research.

18. In the words of one researcher, _____ permit(s) access to "people's ideas, thoughts, and memories in their own words rather than in the words of the researcher." (p. 71)
a.   structured interviews
b.   unstructured interviews
c.   quantitative research
d.   survey research

19. Researchers may use multiple methods of social research in order to: (p. 73)
a.   extend the research over a longer period of time.
b.   determine the one best research method for examining a particular topic.
c.   validate or refine one type of data by use of another type.
d.   cut the overall cost of their study.

20. According to your text, important ethical concerns were raised by all of the following researchers, except: (p. 73)
a.   William Zellner
b.   Laud Humphreys
c.   Rik Scarce
d.   Elijah Anderson

TRUE-FALSE QUESTIONS

T   F   1.   Sociological research indicates that most people who threaten suicide will not commit suicide. (p. 38)

T   F   2.   Descriptive studies attempt to explain cause-and-effect relationships. (p. 44)

T   F   3.   Most sociological studies on suicide have used quantitative research. (p. 47)

T   F   4.   Emile Durkheim was one of the first sociologists to study suicide systematically. (p. 47)

T F   5.   Ratios are the number of times a given event occurs in a population. (p. 50)

T F   6.   Researchers have difficulty making global comparisons regarding certain kinds of behavior, such as suicide. (p. 52)

T F   7.   According to Emile Durkheim, egoistic suicide occurs among people who are isolated from any social group. (p. 56)

T F   8.   Qualitative and quantitative analyses both start with the same set of assumptions. (p. 58)

T F   9.   Experiments, surveys, and analyses of existing data are most frequently used in quantitative research. (p. 59)

T F   10.   The Hawthorne effect refers to changes in the subject's behavior caused by the researcher's presence or by the subject's awareness of being studied. (p. 62)

T F   11.   Computer-assisted telephone interviewing has become much easier to conduct in recent years. (p. 64)

T F   12.   Through an analysis of existing statistics, sociologist K.D. Breault concluded that recent data supported Durkheim's views regarding the relationship between social integration and suicide. (p. 66)

T F   13.   Ethnographic studies often are conducted over a period of years. (p. 69)

T F   14.   Social scientists who believe that quantitative research methods provide the most accurate means of measuring attitudes, beliefs, and behavior often praise the data obtained through field research. (p. 71)

T F   15.   Research is the "life blood" of sociology. (p. 75)

SOCIOLOGY IN OUR TIMES: DIVERSITY ISSUES

1.   Dr. Jack Kevorkian, who was charged with helping two seriously ill women commit suicide in 1991, arrived at the courthouse in Pontiac, Michigan, dressed in cardboard stocks and declared that the case against

him was a throwback to the Middle Ages (*New York Times*, September 15, 1995:A7). Do you agree with his assertion? Why or why not?

2. Does recent research support the finding that women "attempt" suicide more often than men due to problems in their personal relationships, such as being discarded by a lover or husband? Why is men's suicidal behavior often attributed to performance, such as when their self-esteem and independence are threatened?

3. How did sociologist Kevin E. Early attempt to show that "the black church's influence is an essential factor in ameliorating and buffering social forces that otherwise would lead to suicide?"

4. When you examine your textbooks, do you see pictures of people who are similar to you in gender, race/ethnicity, age, class, and ability/disability? Do you think it is important for young children to see textbook pictures of children similar to themselves? Why or why not?

CHAPTER TWO CROSSWORD PUZZLE

For those who enjoy crossword puzzles, here is a puzzle that contains words and names from Chapter Two. Working the puzzle will help you in reviewing the chapter. The answers appear on page 37.

ACROSS

1. Author of 1897 study of suicide
4. He studied homeless women
6. One of the things the normative approach is based on
9. Those persons about whom we want to be able to draw conclusions
10. Another of the things the normative approach is based on
11. Among frequently used independent variables are ___, sex, race, and ethnicity
12. With quantitative research, the ____ is scientific objectivity
13. _____ studies attempt to describe social reality or provide facts about some group, practice, or event
16. Initials of organization that adopted the Code of Ethics in 1971
18. According to Box 2.1, men are ____ likely to kill themselves than are women
19. See 2 down
21. First step in reading a table: read the _____ .
25. Replication: repetition of the investigation in substantially the ____ way it originally was conducted
26. First name of 1 across
27. Studied single-car crashes that might have been "autocides"
28. To demonstrate cause-and-effect relationships, you must show that a correlation exists between the ___ variables
29. The normative approach ____ things such as religion, customs, and tradition to answer important questions
30. With Anselm Strauss, developed the term "grounded theory"
31. See 7 down

DOWN

1. The _____ variable is assumed to depend on or be caused by the independent variable
2. _____ sample: a selection from a larger population that has the essential characteristics of the [answer to 19 across] population
3. With Myra Ferree, she examined illustrations in introductory sociology textbooks
4. One of the authors of Middletown
5. _____ Yoels was interested in why many students do not participate in class discussions in college
6. _____ effect: changes in the subject's behavior caused by the researcher's presence or the subject's awareness of being studied
7. The answer to 31 across expresses the relationship that one category _____ to another
8. The _____ and research cycle
14. One of the basic patterns for variation of suicide models is the two- ____ model
15. Any concept with measurable traits or characteristics that can change or vary
17. A survey is a ____ in which the researcher gathers facts or attempts to determine the relationships between facts
20. Sociology and everyday ____
22. After analyzing the data, you ____ conclusions and report the findings
23. Sampling _____: the extent to which the sample does not represent the population as a whole
24. Operational definition: an explanation of an abstract concept in _____ of observable features

ANSWERS TO PRACTICE TEST, CHAPTER 2

Answers to Multiple Choice Questions

1. d   According to Emile Durkheim, a high suicide rate is symptomatic of a lack of cohesiveness in society. (p. 38)

2. a   The normative approach uses religion, customs, habits, traditions, and law to answer important questions. (p. 43)

3. b   Explanatory studies attempt to explain cause-and-effect relationships and to provide information on why certain events do or do not occur. (p. 44)

4. a   A study of suicidal behavior based on suicide notes is an example of quantitative research. (p. 47)

5. a   The first step in the research process is to select and define the research problem. (p. 47)

6. c   In Emile Durkheim's study of suicide, the degree of social integration was the independent variable. (p. 53)

7. d   This analysis reflects what the text terms multiple causation. (p. 54)

8. b   A random sample refers to a situation in which every member of an entire population has the same chance of being selected for a study. (p. 56)

9. c   Validity is the extent to which a study or research instrument accurately measures what it is supposed to measure; reliability is the extent to which a study or research instrument yields consistent results. (p. 56)

10. d   Researchers following a qualitative approach to research attempt to formulate questions of concern and interest to the "subjects." (p. 58)

11. b   In an experiment, the subjects in the control group are not exposed to the independent variable. (p. 59)

12. a   Suppose that the association of two variables is actually caused by a third variable, but a researcher assumes that the first variable causes the second. This illustrates a spurious correlation. (p. 61)

13. c   An important strength of self-administered questionnaires is low cost. (p. 63)

14. c   Researchers who use existing material and analyze data that originally collected by others are engaged in secondary analysis. (p. 65)

15. b   In the course of a research investigation designed to determine the popularity of exhibits at an art museum, we observe that there is more wear and tear on the floor in front of some exhibits. A content analyst would refer to these patterns as behavioral residues. (p. 66)

16. d   Observation, ethnography, and case studies are examples of field research. (p. 67)

17. a   Taylor's research may best be categorized as participant observation. (p. 68)
18. b   In the words of one researcher, unstructured interviews permit access to "people's ideas, thoughts, and memories in their own words rather than in the words of the researcher." (p. 71)
19. c   Researchers may use multiple methods of social research in order to validate or refine one type of data by use of another type. (p. 73).
20. d   According to your text, important ethical concerns were raised by all of the following researchers, <u>except</u> Elijah Anderson. (p. 73)

Answers to True-False Questions
1. False -- Sociological research indicates that people who threaten to kill themselves often are sending messages to others and may indeed attempt suicide. (p. 38)
2. False -- Explanatory studies attempt to explain cause-and-effect relationships. By contrast, descriptive studies attempt to describe social reality or provide facts about some group, practice, or event. (p. 44)
3. True (p. 47)
4. True (p. 47)
5. False -- Rates are the number of times a given event occurs in a population; ratios express the relationship of one subcategory to another, such as males to females. (p. 50)
6. True (p. 52)
7. True (p. 56)
8. False -- Qualitative analysis often starts with different assumptions than those of quantitative analyses. (p. 58)
9. True (p. 59)
10. True (p. 62)
11. False -- Computer-assisted telephone interviewing has become more difficult to conduct since many people now have telephone answering machines and voice mail, making potential respondents less accessible to researchers. (p. 64)
12. True (p. 66)
13. True (p. 69)
14. False -- Social scientists who believe that quantitative research methods are more accurate than qualitative research methods argue that what is learned about a specific group or community through field research cannot be generalized to a larger population. (p. 71)
15. True (p. 75)

ANSWER TO CHAPTER TWO CROSSWORD PUZZLE

```
D U R K H E I M   L I E B O W   H A B I T
E     E   A       Y   I     A     E     H
P O P U L A T I O N   L A W     A G E   O
E     R   L       D   G O A L   T     R   O
N     E           I         T H     S     R
D E S C R I P T I V E   A S A     O       Y
E     E     E     A   P       M O R E
N     N     A     R   O       N
T O T A L   K   T I T L E     E     D
      A   I     A   L     E       R     T
      T   F     B     E R   S A M E
E M I L E   Z E L L N E R   R     W     R
      V         E     T W O       M
U S E S   G L A S E R     R A T I O S
```

CHAPTER 3
CULTURE

BRIEF CHAPTER OUTLINE

CHAPTER SUMMARY

Culture is the knowledge, language, values, customs, and material objects that are passed from person to person and from one generation to the next. At the macrolevel, culture can be a stabilizing force or a source of discord, conflict, and even violence. At the microlevel, culture is essential for individual survival. Sociologists distinguish between **material culture** -- the physical creations of society -- and **nonmaterial culture** -- the abstract or intangible human creations of society (such as **symbols, language, values**, and **norms**). According to the **Sapir-Whorf hypothesis**, language shapes our perception of reality. For example, language may create and reinforce inaccurate perceptions based on gender, race, ethnicity, or other human

38

attributes. While it is assumed that **high culture** appeals primarily to elite audiences, **popular culture** is believed to appeal to members of the middle and working classes. Cultural change and diversity are intertwined. In the United States, diversity is reflected through race, ethnicity, age, sexual orientation, religion, occupation, and so forth. **Culture shock** refers to the anxiety people experience when they encounter cultures radically different from their own. **Ethnocentrism** -- a belief based on the assumption that one's own culture is superior to others -- is counterbalanced by **cultural relativism** -- the belief that the behaviors and customs of a society must be examined within the context of its own culture. As we look toward even more diverse and global cultural patterns in the twenty-first century, it is important to keep our sociological imaginations actively engaged.

LEARNING OBJECTIVES
After reading Chapter 3, you should be able to:
1.   Explain what culture is and describe how it can be both a stabilizing force and a source of conflict in societies.

2.   Describe the importance of culture in determining how people think and act on a daily basis.

3.   Describe the importance of language and explain the Sapir-Whorf hypothesis.

4.   List and briefly explain ten core values in U.S. society.

5.   Contrast ideal and real culture and give examples of each.

6.   State the definition of norms and distinguish between folkways, mores, and laws.

7.   Distinguish between high culture and popular culture and between fads and fashion.

8. State the functionalist and conflict perspectives on popular culture.

9. Distinguish between discovery, invention, and diffusion as means of cultural change. Explain why the rate of cultural change is uneven.

10. Describe subcultures and countercultures; give examples of each.

11. State the definitions for culture shock, ethnocentrism, and cultural relativism, and explain the relationship between these three concepts.

12. Describe the functionalist, conflict, and interactionist perspectives on culture.

KEY TERMS (defined at page number shown and in glossary)

counterculture 109	language 90
cultural imperialism 117	laws 99
cultural lag 105	material culture 85
cultural relativism 113	mores 98
cultural universals 87	nonmaterial culture 85
culture 81	norms 98
culture shock 111	popular culture 99
diffusion 105	real culture 97
discovery 105	sanctions 98
ethnocentrism 112	Sapir-Whorf hypothesis 90
fad 101	subculture 108
false consciousness 114	taboos 98
fashion 101	technology 85
folkways 98	value contradictions 97
ideal culture 97	values 96
invention 105	

KEY PEOPLE (identified at page number shown)

CHAPTER OUTLINE

I.   CULTURE AND SOCIETY

    A.   **Culture** is the knowledge, language, values, customs, and material objects that are passed from person to person and from one generation to the next in a human group or society.

    B.   **Material culture** consists of the physical or tangible creations that members of a society make, use, and share while **nonmaterial culture** consists of the abstract or intangible human creations of society that influence people's behavior.

    C.   According to anthropologist George Murdock, **cultural universals** are customs and practices that occur across all societies. Examples include appearance, activities, social institutions, and customary practices.

II.   COMPONENTS OF CULTURE

    A.   A **symbol** is anything that meaningfully represents something else.

    B.   **Language** is defined as a set of symbols that express ideas and enable people to think and communicate with one another.

        1.   According to the **Sapir-Whorf hypothesis**, language not only expresses our thoughts and perceptions but also influences our perception of reality.

        2.   Language and gender

            a.   Examples of situations in which the English language ignores women include using the masculine gender to refer to human beings in general, and nouns that show the gender of the person we expect in a particular occupation.

            b.   Words have positive connotations when relating to male power, prestige, and leadership; when related to women, they convey negative overtones of weakness, inferiority, and immaturity.

3. Language, race, and ethnicity
   a. Language may create and reinforce our perceptions about race and ethnicity by transmitting preconceived ideas about the superiority of one category of people over another.
   b. The "voice" of verbs may devalue contributions of members of some racial-ethnic groups.
C. **Values** are collective ideas about what is right or wrong, good or bad, and desirable or undesirable in a particular culture.
   1. Ten U.S. core values are:
      a. Individualism;
      b. Achievement and success;
      c. Activity and work;
      d. Science and technology;
      e. Progress and material comfort;
      f. Efficiency and practicality;
      g. Equality;
      h. Morality and humanitarianism;
      i. Freedom and liberty; and
      j. Racism and group superiority.
   2. **Value contradictions** are values that conflict with one another or are mutually exclusive (achieving one makes it difficult to achieve another).
   3. **Ideal culture** refers to the values and standards of behavior that people in a society profess to hold; **real culture** refers to the values and standards of behavior that people actually follow.
D. **Norms** are established rules of behavior or standards of conduct.
   1. **Folkways** are everyday customs that may be violated without serious consequences within a particular culture.
   2. **Mores** are strongly held norms that may not be violated without serious consequences within a particular culture. **Taboos** are mores so strong that their violation is considered to be extremely offensive.
   3. **Laws** are formal, standardized norms that have been enacted by legislatures and are enforced by formal sanctions.

III. POPULAR CULTURE
   A. High culture consists of activities usually patronized by elite audiences while **popular culture** consists of activities,

42

products, and services which are assumed to appeal primarily to members of the middle and working class.

    B.   **Culture capital theory** is based on the assumption that high culture is a device used by the dominant class to exclude the subordinate classes.

    C.   A **fad** is a temporary but widely copied activity followed enthusiastically by large numbers of people.

    D.   A **fashion** is a currently valued style of behavior, thinking, or appearance that is longer lasting and more widespread than a fad.

    E.   Divergent perspectives on popular culture

        1.   Functionalist theorists suggest that popular culture may be the most widely shared aspect of culture (the "glue") that holds society together.

        2.   Conflict theorists note that popular culture has been turned into a commodity.

IV.   CULTURAL CHANGE AND DIVERSITY

    A.   Cultural change is continual in societies, and these changes are often set in motion by three processes:

        1.   **Discovery** is the process of learning about something previously unknown or unrecognized.

        2.   **Invention** is the process of combining existing cultural items into a new form.

        3.   **Diffusion** is the transmission of cultural items or social practices from one group or society to another.

    B.   According to William F. Ogburn, **cultural lag** is a gap between the technical development (material culture) of a society and its moral and legal institutions (nonmaterial culture).

    C.   Cultural diversity

        1.   A **subculture** is a group of people who share a distinctive set of cultural beliefs and behaviors that differ in some significant way from that of the larger society. Examples include Old Order Amish and Chinatowns.

        2.   A **counterculture** is a group that strongly rejects dominant societal values and norms and seeks alternative lifestyles. Examples include skinheads and members of some paramilitary militias.

    D.   **Culture shock** is the disruption that people feel when they encounter cultures radically different from their own, and they believe they cannot depend on their own taken-for-granted assumptions about life.

    E.   **Ethnocentrism** is the assumption that one's own culture and way of life are superior to all others. By contrast,

    cultural relativism is the belief that the behaviors and customs of a society must be viewed and analyzed within the context of its own culture.

V. SOCIOLOGICAL ANALYSIS OF CULTURE
  A. According to functionalist theorists, societies where people share a common language and core values are more likely to have consensus and harmony.
  B. Conflict theorists suggest that values and norms help create and sustain the privileged position of the powerful in society. According to Karl Marx, people are not aware that they are being dominated because they have **false consciousness**.
  C. According to interactionist theorists, people create, maintain, and modify culture as they go about their everyday activities.

VI. CULTURAL PATTERNS FOR THE TWENTY-FIRST CENTURY
  A. Some of the most important changes in cultural patterns may include: (1) television and radio, films and videos, and electronic communications will continue to accelerate the flow of information; however, (2) most of the world's population will not participate in this technological revolution.
  B. A single global culture is a worldwide interconnection of culture without regard for national identities or boundaries; however, some critics attribute global cultural changes to **cultural imperialism** -- the extensive infusion of one nation's culture into other nations.
  C. The study of culture helps us understand not only our own "tool kit" of symbols, stories, rituals, and world-views, but to expand our insights to include those of other people of the world who also seek strategies for enhancing their own quantity and quality of life.

ANALYZING AND UNDERSTANDING THE BOXES

  After reading the chapter and studying the outline, re-read the four boxes and write down key points and possible questions for class discussion.

Sociology and Everyday Life -- How Much Do You Know About Culture and Hate Crimes?

Key Points:

Discussion Questions:

1.

2.

3.

Sociology and Law -- Dealing with Hate Crimes

Key Points:

Discussion Questions:

1.

2.

3.

Sociology and Media -- Popular Culture, Rap, and Social Protest

Key Points:

Discussion Questions:

1.

2.

3.

Sociology in Global Perspective -- Hostility Toward Immigrants

Key Points:

Discussion Questions:

1.

2.

3.

PRACTICE TEST

MULTIPLE CHOICE QUESTIONS

Select the response that best answers the question or completes the statement:

1. _____ consists of knowledge, language, values, customs, and material objects. (p. 81)
    a. Social structure
    b. Society
    c. Culture
    d. Social organization

2. All of the following statements regarding culture are true, <u>except</u>: (p. 82)
    a. Culture is essential for our survival.
    b. Culture is essential for our communications with other people.
    c. Culture is fundamental for the survival of societies.
    d. Culture is always a stabilizing force for societies.

3. According to the functionalist perspective, cultural universals are: (p. 87-88)
    a. useful because they ensure the smooth and continued operation of society.
    b. the result of attempts by a dominant group to impose its will on a subordinate group.
    c. independent from functional necessities.
    d. very similar in form from one group to another and from one time to another within the same group.

4. The color of baby clothing is an example of: (p. 90)
    a. the use of language to construct social reality.
    b. how symbols affect our thoughts about gender.
    c. cultural universals.
    d. ideal cultural norms.

5.   Regarding the relationship between language and gender, the text points out that: (p. 91)
    a.   the pronouns he and she are seldom used in everyday conversation by most people.
    b.   the English language largely has been purged of sexist connotations.
    c.   the English language ignores women by using the masculine form to refer to human beings in general.
    d.   words in the English language typically have positive connotations when relating to female power, prestige, and leadership.

6.   The most frequently spoken language in U.S. homes other than English is: (p. 94)
    a.   Italian.
    b.   Spanish.
    c.   French.
    d.   German.

7.   From a _____ perspective, a shared language is essential to a common culture. (p. 95)
    a.   functionalist
    b.   conflict
    c.   feminist
    d.   interactionist

8.   Which of the following hypothetical statements does not express a core U.S. value? (pp. 96-97)
    a.   "How well does it work?"
    b.   "Is this a realistic thing to do?"
    c.   "My freedom is important to me."
    d.   "It is good to be lazy."

9.   All of the following are examples of U.S. folkways, except: (p. 98)
    a.   using underarm deodorant.
    b.   brushing our teeth.
    c.   avoiding sexual relationships with siblings.
    d.   wearing appropriate clothing for specific occasions.

10.   The most common type of formal norms is: (98)
    a.   folkways.
    b.   mores.
    c.   sanctions.
    d.   laws.

11. According to the text, all of the following are examples of popular culture, except: (99)
   a. live theater.
   b. spectator sports.
   c. movies.
   d. rock concerts.

12. Sociologist David Halle studied artwork in U.S. homes and determined that _____ were a fashion that crossed all class and racial lines. (p. 101)
   a. portrayals of farm animals
   b. landscape depictions
   c. depictions of children playing
   d. depictions of city life

13. According to functionalist analysts, popular culture may: (102)
   a. reinforce core cultural values.
   b. promote consumption of commodities.
   c. be the "glue" that holds society together.
   d. make us more aware of the problems we face in everyday life.

14. According to the text's discussion of popular culture, rap music: (p. 103)
   a. lacks social creativity.
   b. sometimes advocates violence, exploitation of women, and hatred of the police.
   c. has no positive attributes.
   d. originated in the Los Angeles riots in the 1990s.

15. _____ is the process of reshaping existing cultural items into a new form. (p. 105)
   a. Discovery
   b. Diffusion
   c. Invention
   d. Restoration

16. According to the text, all of the following are examples of countercultures, except: (pp. 109-110)
   a. the Old Order Amish.
   b. neo-Nazi skinheads.
   c. beatniks and flower children.
   d. the Nation of Islam.

17. The disorientation that people feel when they encounter cultures radically different from their own is referred as: (p. 111)
   a. cultural diffusion.
   b. cultural relativism.
   c. cultural disorientation.
   d. cultural shock.

18. International sports competition, such as the Olympic Games, may help to foster: (p. 112)
   a. ethnocentrism.
   b. cultural relativism.
   c. cultural diffusion.
   d. cultural indifference.

19. Anthropologist Marvin Harris has pointed out that the Hindu taboo against killing cattle is very important to the economic system in India. This exemplifies: (p. 113)
   a. ethnocentrism.
   b. cultural relativism.
   c. cultural diffusion.
   d. cultural indifference.

20. Conflict theorists point out that most people are not aware that they are being dominated by others because they believe that their best interests are being served by the members of powerful groups. This human tendency illustrates: (p. 114)
   a. false consciousness.
   b. group consciousness.
   c. cultural relativism.
   d. ethnocentrism.

TRUE-FALSE QUESTIONS

T   F   1.   Hatred and intolerance in a society may be the downside of some "positive" cultural values. (p. 82)

T   F   2.   Humans have a number of basic instincts. (p. 85)

T   F   3.   Technology makes it possible for people to transform resources into usable forms. (p. 85)

T   F   4.   The subject matter of jokes is thought to be a cultural universal. (p. 87)

49

T	F	5.	As a result of their symbolic significance, flags can be a source of discord and strife among people. (p. 88)
T	F	6.	According to the Sapir-Whorf hypothesis, language shapes the view of reality of its speakers. (p. 92)
T	F	7.	Functionalist theorists view language as a source of power and social control that perpetuates inequalities in society. (p. 96)
T	F	8.	According to sociologist Robin M. Williams, racism and group superiority is a core value in the United States. (pp. 96-97)
T	F	9.	Unlike folkways, mores have strong moral and ethical connotations that may not be violated without serious consequences. (p. 98)
T	F	10.	State governments have been unwilling to enact laws intended to discourage or punish prejudice-motivated crimes. (p. 100)
T	F	11.	Fads tend to be longer lasting and more widespread than fashions. (p. 101)
T	F	12.	Cultural lag refers to a gap between the technical development of a society and its moral and legal institutions. (p. 105)
T	F	13.	Chinatowns and other ethnic subcultures help first-generation immigrants adapt to abrupt cultural change. (p. 109)
T	F	14.	According to some interactionist theorists, culture helps people meet their biological, instrumental, and integrative needs. (p. 113)
T	F	15.	Hostility toward immigrants is largely confined to the United States. (p. 115)

SOCIOLOGY IN OUR TIMES: DIVERSITY ISSUES

1. The text states that "how people view culture is intricately related to their location in society with regard to their race/ethnicity, class, sex,

Homework for Oct. 8

and age." (p. 82)   Can you think of examples from your own experiences that either prove or disprove this assertion?

2.   Do you think that recent conflicts over the meaning of symbols such as the Confederate flag reflect racism in the United States or are these conflicts "blown out of proportion" by the proponents and opponents in these controversies?   What other symbols also may generate controversy?

3.   In the 1995 trial of former athlete/sportscaster/actor O. J. Simpson, a detective with the Los Angeles Police Department was widely called a "racist" because evidence showed that he had repeatedly used the derogatory term "nigger" over the past decade. Is there a direct relationship between the use of such derogatory terms and an individual's everyday actions as a police officer?   Why or why not?

4.   How does the "voice" of verbs in these two sentences influence our perceptions about the achievements or activities of people of color:
a.   "African Americans were given the right to vote."
b.   "Columbus discovered America."

5.   What, if anything, do popular cultural images such as Aunt Jemima and Uncle Ben tell us about the relationship between race, gender, and popular culture?   What other popular cultural images have you seen that convey derogatory messages about race/ethnicity, gender, class, age, and ability/disability?

CHAPTER THREE CROSSWORD PUZZLE

For those who enjoy crossword puzzles, here is a puzzle that contains words and names from Chapter Three. Working the puzzle will help you in reviewing the chapter. The answers appear on page 55.

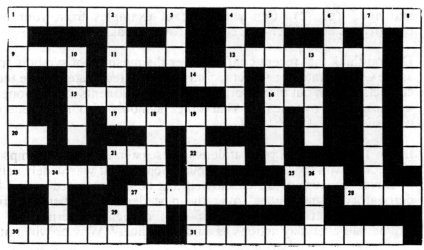

ACROSS

1. Process of learning about something previously unknown or unrecognized
4. Includes the knowledge, techniques, and tools that make it possible to transform resources into usable forms
9. Temporary but widely copied activity
11. Formal, standardized norms enacted by legislatures
12. Whorf's first name
14. Ethnocentrism and xenocentrism are _____ assumptions about comparing cultures
15. The tribe in 8 Down is a _____-industrial economy
16. Cultural _____: a gap between technical development of a society and its moral and legal institutions
17. Language is a set of _____ that express ideas, etc.
20. Technology makes _____ possible for people to transform resources into usable form
21. _____ of the following may be a cultural universal: appearance, activities, social institutions, and customary practices
22. Violating a taboo could be considered "really _____" conduct
23. Established rules of behavior or standards of conduct
25. Middle word in the second core value listed in the text
27. His cultural capital theory views high culture as a device used by the dominant class to exclude the subordinate classes
28. Those who are ethnocentric think of their own culture as the _____
30. The knowledge, language, etc., that are passed from one person to another in a human group or society
31. The Old Order Amish are an example

DOWN

1. Transmission of cultural items or social practices from one group or society to another
2. Collective ideas of what is right and wrong, etc.
3. Is culture important?
4. Their violation is considered to be extremely offensive and even unmentionable
5. Value contradictions: Values that _____ with one another or are mutually exclusive
6. Invention: reshaping existing cultural items into a new f____
7. Ethnocentrism: assumption that one's _____ and way of life are superior to all others
8. Tribe that Chagnon visited
10. A hypothesis bears his and Whorf's names
13. Some historians _____ that cultural bias is shown by the notion that "Columbus discovered America
18. Mores _____ be violated without serious consequences in a particular culture
19. Sociologist who referred to a disparity as cultural lag (and others bearing his name)
24. _____ culture: the values and standards of behavior that people actually follow
26. Culture is passed from one generation to the _____
29. Next-to-last word in definition of culture

ANSWERS TO PRACTICE TEST, CHAPTER 3

Answers to Multiple Choice Questions

1. c   Culture consists of knowledge, language, values, customs, and material objects. (p. 81)
2. d   All of the following statements regarding culture are true, except: culture is always a stabilizing force for societies. (p. 82)
3. a   According to the functionalist perspective, cultural universals are useful because they ensure the smooth and continued operation of society. (p. 87-88)
4. b   The color of baby clothing is an example of how symbols affect our thoughts about gender. (p. 90)
5. c   Regarding the relationship between language and gender, the text points out that the English language ignores women by using the masculine form to refer to human beings in general. (p. 91)
6. b   The most frequently spoken language in U.S. homes other than English is Spanish. (p. 94)
7. a   From a functionalist perspective, a shared language is essential to a common culture. (p. 95)
8. d   The following hypothetical statement does not express a core U.S. value: "It is good to be lazy." (pp. 96-97)
9. c   All of the following are examples of U.S. folkways, except: avoiding sexual relationships with siblings. (p. 98)
10. d   The most common type of formal norms is laws. (p. 98)
11. a   According to the text, all of the following are examples of popular culture, except: live theater. (p. 99)
12. b   Sociologist David Halle studied artwork in U.S. homes and determined that landscape depictions were a fashion that crossed all class and racial lines. (p. 101)
13. c   According to functionalist analysts, popular culture may be the "glue" that holds society together. (102)
14. b   According to the text's discussion of popular culture, rap music sometimes advocates violence, exploitation of women, and hatred of the police. (p. 103)
15. c   Invention is the process of reshaping existing cultural items into a new form. (p. 105)
16. a   According to the text, all of the following are examples of countercultures, except: the Old Order Amish. (pp. 109-110)
17. d   The disorientation that people feel when they encounter cultures radically different from their own is referred as cultural shock. (p. 111)
18. a   International sports competition, such as the Olympic Games, may help to foster ethnocentrism. (p. 112)

19. b Anthropologist Marvin Harris has pointed out that the Hindu taboo against killing cattle is very important to the economic system in India. This exemplifies cultural relativism. (p. 113)

20. a Conflict theorists point out that most people are not aware that they are being dominated by others because they believe that their best interests are being served by the members of powerful groups. This human tendency illustrates false consciousness. (p. 114)

Answers to True-False Questions:

1. True (p. 82).

2. False -- According to sociologists, humans do not have instincts; what we most often think of as instinctive behavior actually can be attributed to reflexes and drives. (p. 85)

3. True (p. 85)

4. False -- Although telling jokes may be a universal practice, what is considered to be a joke in one society may be an insult in another. (p. 87)

5. True (p. 88)

6. True (pp. 90-91)

7. False -- Conflict theorists view language as a source of power and social control; it perpetuates inequalities between people and between groups because words are used to "keep people in their place." (p. 96)

8. True (pp. 96-97)

9. True (p. 98)

10. False -- A number of states have enacted laws defining hate crimes and substantially increasing the punishment for the offense if it is found to be motivated by hate. (p. 100)

11. False -- Fashions tend to be longer lasting and more widespread than fads, which are temporary but widely copied activities followed enthusiastically by large numbers of people. (p. 101)

12. True (p. 105)

13. True (p. 109)

14. False -- Functionalist theorists, such as anthropologist Bronislaw Malinowski, are the ones who suggest that culture helps people meet their biological, instrumental, and integrative needs. (p. 113)

15. False -- According to Box 3.4, hostility toward immigrants has been increasing in a number of countries, including the nations of Western Europe; some people think that immigrant workers are competing for scarce jobs in tough economic times and draining the welfare, education, and health care systems. (p. 115)

ANSWER TO CHAPTER THREE CROSSWORD PUZZLE

```
D  I  S  C  O  V  E  R  Y  ▮  ▮  T  E  C  H  N  O  L  O  G  Y
I  ▮  ▮  ▮  ▮  A  ▮  E  ▮  ▮  ▮  T  ▮  O  ▮  ▮  R  ▮  W  ▮  A
F  A  D  S  ▮  L  A  W  S  ▮  B  E  N  J  A  M  I  N  ▮  A  N
F  ▮  ▮  A  ▮  U  ▮  ▮  T  W  O  ▮  F  ▮  R  ▮  ▮  C  ▮  O  M
U  ▮  ▮  P  R  E  ▮  ▮  ▮  ▮  O  ▮  L  A  G  ▮  ▮  U  ▮  ▮  M
S  ▮  ▮  I  ▮  S  Y  M  B  O  L  S  ▮  I  ▮  U  ▮  ▮  L  ▮  A
I  T  ▮  R  ▮  ▮  A  ▮  G  ▮  ▮  ▮  C  ▮  E  ▮  ▮  T  ▮  M  O
O  ▮  ▮  ▮  ▮  A  N  Y  ▮  B  A  D  ▮  T  ▮  ▮  ▮  U  ▮  ▮  O
N  O  R  M  S  ▮  ▮  N  ▮  U  ▮  ▮  ▮  A  N  D  ▮  R  ▮  ▮  ▮
▮  ▮  E  ▮  ▮  B  O  U  R  D  I  E  U  ▮  E  ▮  B  E  S  T  ▮
▮  ▮  A  ▮  ▮  O  ▮  T  ▮  N  ▮  ▮  ▮  X  ▮  ▮  ▮  ▮  ▮  ▮
C  U  L  T  U  R  E  ▮  ▮  S  U  B  C  U  L  T  U  R  E  S
```

CHAPTER 4
SOCIALIZATION

BRIEF CHAPTER OUTLINE
Why Is Socialization Important?
    Human Development: Biology and Society
    Social Isolation
    Child Maltreatment
Socialization and the Self
    Sociological Theories of Human Development
    Psychological Theories of Human Development
Agents of Socialization
    The Family
    The School
    Peer Groups
    Mass Media
Gender and Racial-Ethnic Socialization
Socialization Through the Life Course
    Infancy and Childhood
    Adolescence
    Adulthood
Resocialization
    Voluntary Resocialization
    Involuntary Resocialization
Socialization in the Twenty-First Century

CHAPTER SUMMARY
Socialization is the lifelong process through which individuals acquire a self-identity and the physical, mental, and social skills needed for survival in society. Socialization is essential for the individual's survival and for human development; it also is essential for the survival and stability of society. People are a product of two forces: heredity and social environment. Most sociologists agree that while biology dictates our physical makeup, the social environment largely determines how we develop and behave. Humans need social contact to develop properly. Cases of feral and isolated children have shown that people who are isolated during their formative years fail to develop their full emotional and intellectual capacities and that social contact is essential in developing a self, or **self-concept**. Charles Horton Cooley developed the image of the **looking-glass self** to explain how people see themselves through the perceptions of others. George Herbert Mead linked the idea of self-concept to **role taking** and to learning the rules of social interaction. When children do not have a positive environment in which to

develop a self-concept, it becomes difficult to form a healthy social self. While Cooley's and Mead's theories are sociologically based, the theories of Sigmund Freud, Erik Erikson, Jean Piaget, Lawrence Kohlberg, and Carol Gilligan are more psychologically based. According to sociologists, **agents of socialization** -- including families, schools, peer groups, the media, and workplace -- teach us what we need to know in order to participate in society. Social class, gender, and race are all determining factors in the life-long socialization process. We learn knowledge and skills for future roles through **anticipatory socialization**. **Resocialization** -- the process of learning new attitudes, values, and behaviors, either voluntarily or involuntarily -- sometimes takes place in **total institutions**. As we approach the twenty-first century, we must not only learn about the past but also acquire the knowledge and skills to think about the future in a practical manner.

LEARNING OBJECTIVES
After reading Chapter 4, you should be able to:

1. Define socialization and explain why this process is essential for the individual and society.

2. Distinguish between sociological and sociobiological perspectives on the development of human behavior.

3. Explain why cases of isolated children are important to our understanding of the socialization process.

4. Explain the key components of the human development theories of Charles Horton Cooley and George Herbert Mead and evaluate the contribution of each to our understanding of the socialization process.

5. Describe Mead's concept of the generalized other and explain why socialization is a two-way process.

6. Explain Freud's views on the conflict between individual desires and the demands of society.

57

7.   Present the stages of psychosocial development proposed by Erik H. Erikson and note the major strengths and weaknesses of his approach.

8.   Outline the stages of cognitive development as set forth by Jean Piaget.

9.   Compare and contrast the moral development theories of Lawrence Kohlberg and Carol Gilligan.

10. State the major agents of socialization and describe their effects on children's development.

11. Explain what is meant by gender socialization and racial socialization.

12. Outline the stages of the life course and explain how each stage varies based on gender, race/ethnicity, class, and positive or negative treatment.

13. Describe the process of resocialization and explain why it often takes place in a total institution.

KEY TERMS (defined at page number shown and in glossary)

agents of socialization 142
anticipatory socialization 150
catharsis theory 146
ego 137
gender socialization 148
generalized other 134
id 137
looking-glass self 132
observational learning theory 146
peer group 144

racial socialization 150
resocialization 153
role-taking 132
self-concept 130
significant other 133
social devaluation 153
socialization 124
sociobiology 127
superego 137
total institution 154

KEY PEOPLE (identified at page number shown)

Charles Horton Cooley 132
Erik H. Erikson 137
Sigmund Freud 137
Carol Gilligan 141
Lawrence Kohlberg 140

George Herbert Mead 132
Wilbert Moore 153
Jean Piaget 139
Edward O. Wilson 127

CHAPTER OUTLINE

I. WHY IS SOCIALIZATION IMPORTANT?

  A. **Socialization** is the lifelong process of social interaction through which individuals acquire a self-identity and the physical, mental, and social skills needed for survival in society.

  B. Human Development: Biology and Society

   1. Every human being is a product of biology, society, and personal experiences, or heredity and environment.

   2. **Sociobiology** is the systematic study of how biology affects social behavior.

  C. Social Isolation

   1. Social environment is a crucial part of an individual's socialization; people need social contact with others in order to develop properly.

   2. Researchers have attempted to demonstrate the effects of social isolation on nonhuman primates that are raised without contact with others of their own species.

   3. Feral children and isolated children illustrate the importance of socialization.

   4. The most frequent form of child maltreatment is child neglect.

II.   SOCIALIZATION AND THE SELF
   A.   Without social contact, we cannot form a **self-concept** -- the totality of our beliefs and feelings about ourselves.
   B.   Sociological Theories of Human Development: Cooley and Mead
      1.   According to **Charles Horton Cooley's looking-glass self,** a person's sense of self is derived from the perceptions of others through a three step process:
         a.   We imagine how our personality and appearance will look to other people.
         b.   We imagine how other people judge the appearance and personality that we think we present.
         c.   We develop a self-concept.
      2.   **George Herbert Mead** linked the idea of self concept to **role-taking** -- the process by which a person mentally assumes the role of another person in order to understand the world from that person's point of view.
         a.   **Significant others** are those persons whose care, affection, and approval are especially desired and who are most important in the development of the self; these individuals are extremely important in the socialization process.
         b.   Mead divided the self into the "I" -- the subjective element of the self that represents the spontaneous and unique traits of each person -- and the "me" -- the objective element of the self, which is composed of the internalized attitudes and demands of other members of society and the individual's awareness of those demands.
         c.   Mead outlined three stages of self development:
            (1) preparatory stage -- children largely imitate the people around them;
            (2) play stage (from about age 3 to 5) -- children learn to use language and other symbols, thus making it possible for them to pretend to take the roles of specific people;
            (3) game stage -- children understand not only their own social position but also the positions of others around them. At this time, the child develops a **generalized other** -- an awareness

of the demands and expectations of the society as a whole or of the child's subculture.

3. Interactionist perspectives such as Cooley's and Mead's contribute to our understanding of how the self develops; however, these theories often do not take into account differences in people's experiences based on race/ethnicity, class, religion, gender, or other factors.

C. Psychological Theories of Human Development

1. In his psychoanalytic perspective, Sigmund Freud divided the mind into three interrelated parts:

   a. The **id** is the component of personality that includes all of the individual's basic biological drives and needs that demand immediate gratification.

   b. The **ego** is the rational, reality-oriented component of personality that imposes restrictions on the innate pleasure-seeking drives of the id.

   c. The **superego** consists of the moral and ethical aspects of personality. When a person is well-adjusted, the ego successfully manages the opposing forces of the id and the superego.

2. **Erik H. Erikson** identified eight psychosocial stages of development, each of which is accompanied by a crisis or potential crisis that involves transitions in social relationships:

   a. Trust versus Mistrust (birth to age 1).

   b. Autonomy versus Shame and Doubt (1-3 years).

   c. Initiative versus Guilt (3-5 years).

   d. Industry versus Inferiority (6-11 years).

   e. Identity versus Role Confusion (12-18 years).

   f. Intimacy versus Isolation (18-35 years).

   g. Generativity versus Self-absorption (35-55 years).

   h. Integrity versus Despair (maturity and old age).

3. **Jean Piaget's** theory of cognitive development is based on the assumption that there are four stages of cognitive development based on how children understand the world around them:

   a. Sensorimotor Stage (birth to age 2) -- children understand the world only though sensory contact and immediate action because they cannot engage in symbolic thought or use language.

   b. Preoperational Stage (ages 2-7) -- children begin to use words as mental symbols and to develop the ability to use mental images.

   c. Concrete Operational Stage (ages 7-11) -- children think in terms of tangible objects and actual events; they also can draw conclusions about the likely physical consequences of an action without always having to try it out.

   d. Formal Operational Stage (age 12 through adolescence) -- adolescents are able to engage in highly abstract thought and understand places, things, and events they have never seen. Beyond this point, changes in thinking are a matter of changes in degree rather than in the nature of their thinking.

4. **Lawrence Kohlberg** set forth three levels of moral development:

   a. Preconventional Level (ages 7-10) -- children give little consideration to the views of others.

   b. Conventional Level (age 10 through adulthood) -- children initially believe that behavior is right if it receives wide approval from significant others, including peers, and then a law-and-order orientation, based on how one conforms to rules and laws.

   c. Postconventional Level (few adults reach this stage) -- people view morality in terms of individual rights. At the final stage of moral development, "moral conduct" is judged by principles based on human rights that transcend government and laws.

5. Gender and Moral Development

   a. One of the major critics of Kohlberg's work was psychologist **Carol Gilligan**, who noted that Kohlberg's model was based solely on male responses.

   b. To correct this perceived oversight, Gilligan examined morality in women by interviewing twenty-eight pregnant women who were deciding whether or not to have an abortion.

62

III.   AGENTS OF SOCIALIZATION
   A.   **Agents of socialization** are the persons, groups, or institutions that teach us what we need to know in order to participate in society. These are the most pervasive agents of socialization in childhood:
   B.   The **family** is the most important agent of socialization in all societies.
      1.   Functionalists emphasize that families are the primary locus for the procreation and socialization of children, as well as the primary source of emotional support.
      2.   To a large extent, the family is where we acquire our specific social position in society.
      3.   Conflict theorists stress that socialization reproduces the class structure in the next generation.
   C.   The school has played an increasingly important role in the socialization process as the amount of specialized technical and scientific knowledge has expanded rapidly.
      1.   Schools teach specific knowledge and skills; they also have a profound effect on a child's self-image, beliefs, and values.
      2.   From a functionalist perspective, schools are responsible for: (1) socialization -- teaching students to be productive members of society; (2) transmission of culture; (3) social control and personal development; and (4) the selection, training, and placement of individuals on different rungs in the society.
      3.   According to conflict theorists such as Samuel Bowles and Herbert Gintis, much of what happens in school amounts to a hidden curriculum -- the process by which children from working class and lower income families learn to be neat, to be on time, to be quiet, to wait their turn, and to remain attentive to their work -- attributes that are important for later roles in the work force.
   D.   A **peer group** is a group of people who are linked by common interests, equal social position, and (usually) similar age.
      1.   Peer groups function as agents of socialization by contributing to our sense of "belonging" and our feelings of self worth.
      2.   Individuals must earn their acceptance with their peers by meeting the group's demands for a high level of conformity to its own norms, attitudes, speech, and dress codes.

E. The **mass media** is an agent of socialization that has a profound impact on both children and adults.
1. The media function as socializing agents in several ways: (1) they inform us about events; (2) they introduce us to a wide variety of people; (3) they provide an array of viewpoints on current issues; (4) they make us aware of products and services that, if we purchase them, supposedly will help us to be accepted by others; and (5) they entertain us by providing the opportunity to live vicariously (through other people's experiences).
2. Television is the most pervasive form of media; ninety-eight percent of all homes in the United States have at least one television set.
    a. Television has been blamed for its potentially harmful effects, such as the declining rate of literacy, rampant consumerism, and increases in violent crime.
    b. Two theories have been most widely used to explain how televised consumerism and violence affect children's behavior: (1) **observational learning theory** states that we observe the behavior of another person and repeat the behavior ourselves, and (2) **catharsis theory** states that televised materialism and violence provide viewers a vicarious outlet for their own greed or aggressiveness.

IV. GENDER AND RACE/ETHNIC SOCIALIZATION
A. **Gender socialization** is the aspect of socialization that contains specific messages and practices concerning the nature of being female or male in a specific group or society.
1. Families, schools, and sports tend to reinforce traditional roles through gender socialization.
2. From an early age, media, including children's books, television programs, movies, and music provide subtle and not-so subtle messages about masculine and feminine behavior.
B. **Racial socialization** is the aspect of socialization that contains specific messages and practices concerning the nature of one's racial or ethnic status as it relates to: (1) personal and group identity; (2) intergroup and inter-individual relationships, and (3) position in the social hierarchy.

V. SOCIALIZATION THROUGH THE LIFE COURSE

A. Socialization is a lifelong process: each time we experience a change in status, we learn a new set of rules, roles, and relationships.

   1. Even before we enter a new status, we often participate in **anticipatory socialization** -- the process by which knowledge and skills are learned for future roles.

   2. The most common categories of age are infancy, childhood, adolescence, and adulthood (often subdivided into young adulthood, middle adulthood, and older adulthood).

B. During infancy and early childhood, family support and guidance are crucial to a child's developing self-concept; however, some families reflect the discrepancy between cultural ideals and reality -- children grow up in a setting characterized by fear, danger, and risks that are created by parental neglect, emotional maltreatment, or premature economic and sexual demands.

C. Anticipatory socialization for adult roles often is associated with adolescence; however, some young people may plunge into adult responsibilities at this time.

D. In early adulthood (usually until about age forty), people work toward their own goals of creating meaningful relationships with others, finding employment, and seeking personal fulfillment. **Wilbert Moore** divided workplace, or occupational, socialization into four phases:

   1. career choice;
   2. anticipatory socialization;
   3. conditioning and commitment; and
   4. continuous commitment.

E. Between the ages of 40 and 60, people enter middle adulthood, and many begin to compare their accomplishments with their earlier expectations.

F. In older adulthood, some people are quite happy and content; others are not. Difficult changes in adult attitudes and behavior may occur in the last years of life when people experience decreased physical ability and **social devaluation** -- when a person or group is considered to have less social value than other groups.

G. It is important to note that everyone does not go through certain passages or stages of a life course at the same age and that life course patterns are strongly influenced by race, ethnicity, and social class, as well.

65

VI.   RESOCIALIZATION
    A.   **Resocialization**   is the process of learning a new and different set of attitudes, values, and behaviors from those in one's previous background and experience.
        1.   **Voluntary resocialization** occurs when we enter a new status of our own free will (e.g., medical or psychological treatment or religious conversion).
        2.   **Involuntary resocialization** occurs against a person's wishes and generally takes place within a **total institution** -- a place where people are isolated from the rest of society for a set period of time and come under the control of the officials who run the institution. Examples include military boot camps, prisons, concentration camps, and some mental hospitals.
VII.   SOCIALIZATION IN THE TWENTY-FIRST CENTURY
    A.   Families are likely to remain the institution that most fundamentally shapes and nurtures personal values and self-identity.
    B.   However, parents increasingly may feel overburdened by this responsibility, especially without societal support -- such as high-quality, affordable child care -- and more education in parenting skills.
    C.   A central issue facing parents and teachers as they socialize children is the growing dominance of the media and other forms of technology.
    D.   In the twenty-first century, socialization must anticipate -- and consider the consequences of -- the future.

ANALYZING AND UNDERSTANDING THE BOXES
        After reading the chapter and studying the outline, re-read the four boxes and write down key points and possible questions for class discussion.

Sociology and Everyday Life -- "How Much Do You Know About Child Maltreatment?"

Key Points:

Discussion Questions:

1.

2.

3.

Sociology and Law -- "Child Maltreatment Then and Now"

Key Points:

Discussion Questions:

1.

2.

3.

Sociology and Media -- "Public Awareness of Child Maltreatment"

Key Points:

Discussion Questions:

1.

2.

3.

Sociology in Global Perspective -- "Child Maltreatment in Asia"

Key Points:

Discussion Questions:

1.

2.

3.

PRACTICE TEST

MULTIPLE CHOICE QUESTIONS

Select the response that best answers the question or completes the statement:

1. _____ is the lifelong process of social interaction through which individuals acquire a self-identity. (p. 124)
    a. Human development
    b. Socialization
    c. Behavior modification
    d. Imitation

2. In discussing child maltreatment, the text points out that: (pp. 125-126)
    a. many types of neglect constitute child maltreatment.
    b. the extent of this problem has been exaggerated by the media.
    c. most sexual abuse perpetrators are punished by imprisonment.
    d. most child maltreatment occurs in families living below the poverty line.

3. The systematic study of how biology affects social behavior is known as _____. (p. 127)
    a. sociophysiology.
    b. sociobiology.
    c. sociology.
    d. social psychology.

4. Harry and Margaret Harlow's experiment with rhesus monkeys demonstrated that: (p. 128)
    a. food was more important to the monkeys than warmth, affection, and physical comfort.
    b. monkeys cannot distinguish between a nonliving "mother substitute" and their own mother.
    c. socialization is not important to rhesus monkeys; their behavior is purely instinctive.
    d. without socialization, young monkeys do not learn normal social or emotional behavior.

5.   All of the following are components of the self-concept, <u>except</u>:
     (pp. 130-131)
     a.   the physical self.
     b.   the active self.
     c.   the functional self.
     d.   the psychological self.

6.   The theories of Charles Horton Cooley and George Herbert Mead can
     best be classified as _____ perspectives. (p. 132)
     a.   interactionist
     b.   functionalist
     c.   conflict
     d.   feminist

7.   According to Charles Horton Cooley, we base our perception of who
     we are on how we think other people see us and on whether this
     seems good or bad to us. He referred to this perspective as the:
     (p. 132)
     a.   self-fulfilling prophecy.
     b.   generalized other.
     c.   looking-glass self.
     d.   significant other.

8.   All of the following are stages in Mead's theory of self development,
     <u>except</u> the: (p. 134)
     a.   anticipatory stage.
     b.   game stage.
     c.   play stage.
     d.   preparatory stage.

9.   The _____ refers to the child's awareness of the demands and
     expectations of society as a whole or of the child's subculture.
     (p. 134)
     a.   looking-glass self
     b.   id
     c.   ego
     d.   generalized other

10. According to Sigmund Freud, the _____ consists of the moral and
     ethical aspects of personality. (p. 137)
     a.   id
     b.   ego
     c.   super ego
     d.   libido

11. Which of the following is not one of Jean Piaget's stages of cognitive development? (pp. 139-140)
    a. preoperational
    b. concrete operational
    c. formal operational
    d. post operational

12. The stages of moral development were initially set forth by _____ and then criticized by _____. (pp. 140-141)
    a. Charles Horton Cooley -- George Herbert Mead
    b. Lawrence Kohlberg -- Carol Gilligan
    c. Jean Piaget -- Lawrence Kohlberg
    d. Erik Erikson -- Jean Piaget

13. According to the text, the most important agent of socialization in all societies is the: (p. 142)
    a. family.
    b. peer group.
    c. school.
    d. church.

14. Sociologist Melvin Kohn has suggested that _____ is one of the strongest influences on what and how parents teach their children. (p. 143)
    a. race/ethnicity
    b. religion
    c. social class
    d. age

15. Currently, more than ____ percent of all U.S. preschool children are in day care of one kind or another. (p. 143)
    a. 15
    b. 25
    c. 50
    d. 75

16. As agents of socialization, peer groups are thought to "pressure" children and adolescents because: (p. 145)
    a. individualism is encouraged and rewarded in these groups.
    b. individuals must earn their acceptance with their peers by conforming to the group's norms.
    c. individuals are encouraged to put friendship above material possessions.
    d. individuals are discouraged from making long-term friends in the peer group.

17. _____ theory states that we observe the behavior of another person and repeat the behavior ourselves. (p. 146)
    a.   Observational learning
    b.   Catharsis
    c.   Patterning
    d.   Resocialization

18. The process by which knowledge and skills are learned for future roles is known as: (p. 150)
    a.   resocialization.
    b.   anticipatory socialization.
    c.   cybersocialization.
    d.   expectant socialization.

19. Social devaluation is most likely to be experienced during this stage of the life course: (p. 153)
    a.   infancy and childhood.
    b.   adolescence.
    c.   early adulthood.
    d.   older adulthood.

20. All of the following are examples of voluntary resocialization, except: (p. 154)
    a.   becoming a student.
    b.   going to prison.
    c.   becoming a Buddhist.
    d.   joining Alcoholics Anonymous.

TRUE-FALSE QUESTIONS

T   F   1.   The socialization process is most effective when people conform to the norms of society because they believe this is the best course of action. (p. 125)

T   F   2.   Unlike humans, nonhuman primates such as monkeys and chimpanzees do not need social contact with others of their species in order to develop properly. (p. 128)

T   F   3.   Social scientists generally agree that feral children were raised by wild animals. (p. 129)

T	F	4.	The cases of "Anna" and "Genie" make us aware of the importance of the socialization process because they show the detrimental effects of social isolation and neglect. (p. 130)
T	F	5.	Historically, U.S. children have been viewed as the property of their parents. (p. 131)
T	F	6.	According to George H. Mead, our looking-glass self is based on our perception of how other people think of us. (p. 132)
T	F	7.	Sigmund Freud theorized that our personalities are largely unconscious -- hidden away outside our normal awareness. (pp. 137-138)
T	F	8.	Erik Erikson suggested that, if infants receive good care and nurturing from their parents, they will develop a sense of trust that is important to their future psychosocial development. (p. 137)
T	F	9.	The most pervasive agents of socialization in childhood are the family, the school, and the church. (p. 142)
T	F	10.	Conflict theorists assert that students have different experiences in the school system, depending on their class, racial-ethnic background, gender, and the neighborhood in which they live. (p. 143)
T	F	11.	The pervasive nature of the mass media as an agent of socialization is exemplified by the fact that about 98 percent of all U.S. homes have at least one television set. (p. 146)
T	F	12.	Gender socialization contributes to people's beliefs about what the "preferred" sex of a child should be and in influencing our beliefs about acceptable behaviors for males and females. (p. 148)
T	F	13.	Racial socialization occurs only when we are children and adolescents. (p. 150)

T	F	14.	Occupational socialization tends to be most intense immediately after a person makes the transition from school to the workplace. (p. 153)
T	F	15.	Voluntary socialization generally takes place within a total institution. (p. 154)

SOCIOLOGY IN OUR TIMES: DIVERSITY ISSUES

1. Do you think race/ethnicity, class, religion, and other factors contribute to inherent tensions between the meanings you derive from your personal experiences and those you take from culture? Can you give specific examples?

2. Is what we know about human development limited by the fact that many theorists have limited their research to white, middle-class respondents? Why or why not?

3. Trace gender and racial socialization in your own life. When did you first become aware of your racial-ethnic category? Of gender expectations regarding your appearance and behavior?

CHAPTER FOUR CROSSWORD PUZZLE

For those who enjoy crossword puzzles, here is a puzzle that contains words and names from Chapter Four. Working the puzzle will help you in reviewing the chapter. The answers appear on page 77.

ACROSS

1. _____ socialization: aspect of socialization that contains specific messages and practices concerning the nature of ones racial or ethnic status

2. The part of the mind that consists of the moral and ethical aspects of personality

6. A former colleague of Lawrence Kohlberg, she asserted that his research had key weaknesses

10. We _____ how to walk, talk, eat, etc., from the culture in which we are raised

11. The component of personality that is rational, reality-oriented, and imposes restrictions

13. Resocialization: voluntary __ involuntary

14. An agent of socialization

15. Peers usually are linked by common interests, equal social position, and similar _____

16. The persons, groups, or institutions that teach us what we need to know in order to participate in society

19. Resocialization: learning a new ____ different set of attitudes, values, and behaviors

20. Component of personality that includes basic biological drives and needs

21. Sociobiology: the systematic _____ of how biology affects social behavior

23. Another word for adolescent under many definitions

24. In Mead's "play stage," children may play at taking the _____ of another

25. First name of child found at age 13 who had been locked in a bedroom since she was 20 months old

26. Margaret _____ Harry Harlow

29. Observational learning theory: we observe the behavior of another person and _____ it ourselves

33. Many parents would prefer a _____

34. _____ children are assumed to have been raised by animals in the wilderness

35. To Cooley, here is where we get a sense of self

36. _____ socialization: process by which knowledge and skills are learned for future roles

DOWN

1. Process by which one mentally assumes the role of another to understand the world from that person's view

2. _____ H. Cooley originated the looking-glass self

3. Isolated child who was kept in an attic-like room

4. _____ other: a person whose care, affection, and approval are especially desired

5. Psychologist who identified eight stages of development, each accompanied by a crisis

7. The Harlows demonstrated that social _____ is bad for nonhuman primates

8. Self-concept includes a step where we _____ how others judge our appearance and personality

9. Founder of psychoanalytic theory

12. Sociologist who asserted we base our perception of who we are on how we think others see us

17. _____-identity: our perception about what kind of people we are

18. Theory that televised materialism and violence provide a vicarious outlet for greed or aggressiveness

22. Self-_____: the totality of our beliefs and feelings about ourselves

27. Agent of socialization involving print or electronic means of communication

28. Psychologist who pioneered cognitive development

30. Gender socialization: aspect of socialization _____ contains messages and practices concerning being female or male

31. One of Mead's stages of self-development

32. In a total institution, a person is deprived of _____ personal possessions, family, and friends

ANSWERS TO PRACTICE TEST, CHAPTER 4

Answers to Multiple Choice Questions

1. b   Socialization is the lifelong process of social interaction through which individuals acquire a self-identity. (p. 124)
2. a   In discussing child maltreatment, the text points out that many types of neglect constitute child maltreatment. (pp. 125-126)
3. b   The systematic study of how biology affects social behavior is known as sociobiology. (p. 127)
4. d   Harry and Margaret Harlow's experiment with rhesus monkeys demonstrated that without socialization young monkeys do not learn normal social or emotional behavior. (p. 128)
5. c   All of the following are components of the self-concept, except: the functional self. (pp. 130-131)
6. a   The theories of Charles Horton Cooley and George Herbert Mead can best be classified as interactionist perspectives. (p. 132)
7. c   According to Charles Horton Cooley, we base our perception of who we are on how we think other people see us and on whether this seems good or bad to us. He referred to this perspective as the looking-glass self. (p. 132)
8. a   All of the following are stages in Mead's theory of self development, except the anticipatory stage. (p. 134)
9. d   The generalized other refers to the child's awareness of the demands and expectations of society as a whole or of the child's subculture. (p. 134)
10. c   According to Sigmund Freud, the super ego consists of the moral and ethical aspects of personality. (p. 137)
11. d   The post operational stage is not one of Jean Piaget's stages of cognitive development. (pp. 139-140)
12. b   The stages of moral development were initially set forth by Lawrence Kohlberg and then criticized by Carol Gilligan. (pp. 140-141)
13. a   According to the text, the most important agent of socialization in all societies is the family. (p. 142)
14. c   Sociologist Melvin Kohn has suggested that social class is one of the strongest influences on what and how parents teach their children. (p. 143)
15. c   Currently, more than 50 percent of all U.S. preschool children are in day care of one kind or another. (p. 143)
16. b   As agents of socialization, peer groups are thought to "pressure" children and adolescents because individuals must earn their acceptance with their peers by conforming to the group's norms. (p. 145)

17.   a   Observational learning theory states that we observe the behavior of another person and repeat the behavior ourselves. (p. 146)

18.   b   The process by which knowledge and skills are learned for future roles is known as anticipatory socialization. (p. 150)

19.   d   Social devaluation is most likely to be experienced during older adulthood. (p. 153)

20.   b   All of the following are examples of voluntary resocialization, except: going to prison. (p. 154)

Answer to True-False Questions

1.   True (p. 125)

2.   False -- Even nonhuman primates need social contact with others of their species in order to develop properly. (p. 128)

3.   False -- Social scientists generally agree that it is highly unlikely that feral children actually are raised by wild animals. They suggest that the children likely were abandoned by their parents shortly before they were found by others. (p. 129)

4.   True (p. 130)

5.   True (p. 131)

6.   False -- Charles H. Cooley originated the concept of the looking-glass self. (p. 132)

7.   True (pp. 137-138)

8.   True (p. 137)

9.   False -- The most pervasive agents of socialization in childhood are the family, the school, peer groups, and the mass media. (p. 142)

10.   True (p. 143)

11.   True (p. 146)

12.   True (p. 148)

13.   False -- Like other forms of socialization, we receive racial socialization throughout our lives. (p. 150)

14.   True (p. 153)

15.   False -- Involuntary socialization generally takes place within a total institution. (p. 154)

ANSWER TO CHAPTER FOUR CROSSWORD PUZZLE

R	A	C	I	A	L		S	U	P	E	R	E	G	O		G	I	L	L	I	G	A	N	
O		H		I			I		R				S			M							F	
L	E	A	R	N		E	G	O		I		C			O		A				O	R		
E		R		N			S	C	H	O	O	L		L		A	G	E			I		E	
T		L		I						O		A		I								U		
A	G	E	N	T	S	O	F	S	O	C	I	A	L	I	Z	A	T	I	O	N	A	N	D
K	S			E		I		A		E						E								
I	D		L	C		S	T	U	D	Y				O				C						
N			F	A		H		T	E	E	N				R	O	L	E						
G	E	N	I	E		A	N	D		A		M					N							
			J		T			R	E	P	E	A	T				C				P			
A		M	A	L	E			S		D		H				F	E	R	A	L				
L			A		I			I		I		A					P			A				
L	O	O	K	I	N	G	G	L	A	S	S		A	N	T	I	C	I	P	A	T	O	R	Y

CHAPTER 5
SOCIAL STRUCTURE AND INTERACTION IN EVERYDAY LIFE

BRIEF CHAPTER OUTLINE
Social Structure: The Macrolevel Perspective
Components of Social Structure
> Status
> Role
> Groups
> Social Institutions

Societies: Changes in Social Structure
> Mechanical and Organic Solidarity
> *Gemeinschaft* and *Gesellschaft*
> Social Structure and Homelessness

Social Interaction: The Microlevel Perspective
> Social Interaction and Meaning
> The Social Construction of Reality
> Ethnomethodology
> Dramaturgical Analysis
> The Sociology of Emotions
> Nonverbal Communications

Changing Social Structure and Interaction in the Twenty-first Century

CHAPTER SUMMARY
Social structure and interaction are critical components of everyday life. At the microlevel, **social interaction** -- the process by which people act toward or respond to other people -- is the foundation of meaningful relationships in society. At the macrolevel, **social structure** is the stable pattern of social relationships that exist within a particular group or society. This structure includes social institutions, groups, statuses, roles, and norms. Changes in social structure may dramatically affect individuals and groups, as demonstrated by Durkheim's concepts of **mechanical** and organic solidarity and Tonnies' *Gemeinschaft* and *Gesellschaft*. Social interaction within a society is guided by certain shared meanings of how we behave. Race, ethnicity, gender, and social class often influence perceptions of meaning, however. The **social construction of reality** refers to the process by which our perception of reality is shaped by the subjective meaning we give to an experience. **Ethnomethodology** is the study of the commonsense knowledge that people use to understand the situations in which they find themselves. **Dramaturgical analysis** is the study of social interaction that compares everyday life to a theatrical presentation. **Presentation of self** refers to efforts to present our own self to others in ways that are most favorable to

our own interests or image. Feeling rules shape the appropriate emotions for a given role or specific situation. Social interaction also is marked by **nonverbal communication**, which is the transfer of information between people without the use of speech. As we enter the twenty-first century, macrolevel and microlevel analyses are essential in the determination of how our social structures should be shaped so that they can respond to pressing needs.

LEARNING OBJECTIVES
After reading Chapter 5, you should be able to:
1.   State the definition of social structure and explain why it is important for individuals and society.

2.   State the definition of status and distinguish between ascribed and achieved statuses.

3.   Explain what is meant by master status and give at least three examples.

4.   Define role expectation, role performance, role conflict, and role strain, and give an example of each.

5.   Describe the process of role exiting.

6.   Explain the difference in primary and secondary groups.

7.   Define formal organization and explain why many contemporary organizations are known as "people-processing" organizations.

8. State the definition for social institution and name the major institutions found in contemporary society.

9. Evaluate functionalist and conflict perspectives on the nature and purpose of social institutions.

10. Compare Emile Durkheim's typology of mechanical and organic solidarity with Ferdinand Tonnies' *Gemeinschaft* and *Gesellschaft*.

11. Explain what interactionists mean by the social construction of reality.

12. Describe ethnomethodology and note its strengths and weaknesses.

13. Describe Goffman's dramaturgical analysis and explain what he meant by presentation of self.

14. Explain what is meant by the sociology of emotions and describe sociologist Arlie Hochschild's contribution to this area of study.

15. Define nonverbal communication and personal space and explain how these concepts relate to our interactions with others.

KEY TERMS (defined at page number shown and in glossary)

achieved status 164
ascribed status 164
dramaturgical analysis 180
ethnomethodology 177
formal organization 170
Gemeinschaft 173
Gesellschaft 174
master status 164
mechanical solidarity 173
nonverbal communication 183
organic solidarity 173
personal space 186
presentation of self 180
primary group 169
role 166
role conflict 167
role exit 169

role expectation 166
role performance 166
role strain 167
secondary group 169
self-fulfilling prophecy 177
social construction of reality 176
social group 169
social institution 171
social interaction 159
social marginality 162
social network 170
social structure 160
status 163
status set 164
status symbol 165
stigma 162

KEY PEOPLE (identified at page number shown)

Helen Rose Fuchs Ebaugh 169
Harold Garfinkel 178
Erving Goffman 176
Edward Hall 185
Arlie Hochschild 182

David Snow and
    Leon Anderson 161
Ferdinand Tonnies 173
Jacqueline Wiseman 177

CHAPTER OUTLINE

I.   SOCIAL STRUCTURE: THE MACROLEVEL PERSPECTIVE
    A.   **Social structure** is the stable pattern of social relationships that exist within a particular group or society.
    B.   Social structure creates boundaries that define which persons or groups will be the "insiders" and which will be the "outsiders."
        1.   **Social marginality** is the state of being part insider and part outsider in the social structure. Social marginality results in stigmatization.
        2.   A **stigma** is any physical or social attribute or sign that so devalues a person's social identity that it disqualifies that person from full social acceptance.

II.   COMPONENTS OF SOCIAL STRUCTURE
    A.   A **status** is a socially defined position in a group or society characterized by certain expectations, rights, and duties.
        1.   A **status set** is comprised of all the statuses that a person occupies at a given time.

2.   Ascribed and achieved statuses:
     a.   An **ascribed status** is a social position conferred at birth or received involuntarily later in life. Examples of ascribed statuses include race/ethnicity, age, and gender.
     b.   An **achieved status** is a social position a person assumes voluntarily as a result of personal choice, merit, or direct effort. Examples of achieved statuses include occupation, education and income. Ascribed statuses have a significant influence on the achieved statuses we occupy.
3.   A **master status** is the most important status a person occupies; it dominates all of the individual's other statuses and is the overriding ingredient in determining a person's general social position (e.g., being poor or rich is a master status).
4.   **Status symbols** are material signs that inform others of a person's general social position. Examples include a wedding ring or a Rolls-Royce automobile.

B.   A **role** is a set of behavioral expectations associated with a given status.
1.   **Role expectation** -- a group's or society's definition of the way a specific role ought to be played -- may sharply contrast with **role performance** -- how a person actually plays the role.
2.   **Role conflict** occurs when incompatible role demands are placed on a person by two or more statuses held at the same time (e.g., a woman whose roles include full-time employee, mother, wife, caregiver for an elderly parent, and community volunteer).
3.   **Role strain** occurs when incompatible demands are built into a single status that a person occupies (e.g., a doctor in a public clinic who is responsible for keeping expenditures down and providing high quality patient care simultaneously). Sexual orientation, age, and occupation frequently are associated with role strain.
4.   **Role exit** occurs when people disengage from social roles that have been central to their self-identity (e.g., ex-convicts, ex-nuns, retirees, and divorced women and men).

C.   A **social group** consists of two or more people who interact frequently and share a common identity and a feeling of interdependence.

    1.   A **primary group** is a small, less specialized group in which members engage in face-to-face, emotion-based interactions over an extended period of time (e.g., one's family, close friends, and school or work-related peer groups).

    2.   A **secondary group** is a larger, more specialized group in which the members engage in more impersonal, goal-oriented relationships for a limited period of time (e.g., schools, churches, the military, and corporations).

    3.   A **social network** is a series of social relationships that link an individual to others.

    4.   A **formal organization** is a highly structured group formed for the purpose of completing certain tasks or achieving specific goals (e.g., colleges, corporations, and the government).

D.   A **social institution** is a set of organized beliefs and rules that establish how a society will attempt to meet its basic social needs. Examples of social institutions include the family, religion, education, the economy, the government, mass media, sports, science and medicine, and the military.

III.   SOCIETIES: CHANGES IN SOCIAL STRUCTURE

A.   Sociologists Emile Durkheim and Ferdinand Tonnies developed typologies to explain how stability and change occur in the social structure of societies.

B.   Durkheim's Typology:

    1.   **Mechanical solidarity** refers to the social cohesion in preindustrial societies where there is minimal division of labor and people feel united by shared values and common social bonds.

    2.   **Organic solidarity** refers to the social cohesion found in industrial societies in which people perform very specialized tasks and feel united by their mutual dependence.

C.   Tonnies' Typology:

    1.   According to Ferdinand Tonnies, the *gemeinschaft* is a traditional society in which social relationships are based on personal bonds of friendship and kinship and on intergenerational stability. Relationships are based on ascribed statuses.

    2.   The *gesellschaft* is a large, urban society, in which social bonds are based on impersonal and specialized

relationships, with little long-term commitment to the group or consensus on values. Relationships are based on achieved statuses.

IV.   SOCIAL INTERACTION: THE MICROLEVEL PERSPECTIVE
  A.   Social interaction within a given society has certain shared meanings across situations; however, everyone does not interpret social interaction rituals in the same way.
  B.   The **social construction of reality** is the process by which our perception of reality is shaped largely by the subjective meaning that we give to an experience.
  C.   Our definition of the situation can result in a **self-fulfilling prophecy** -- a false belief or prediction that produces behavior that makes the original false belief come true.
  D.   **Ethnomethodology** is the study of the commonsense knowledge that people use to understand the situations in which they find themselves.
       1.   This approach challenges existing patterns of conventional behavior in order to uncover people's background expectancies, that is, their shared interpretation of objects and events, as well as the actions they take as a result.
       2.   To uncover people's background expectancies, ethnomethodologists frequently conduct breaching experiments in which they break "rules" or act as though they do not understand some basic rule of social life so that they can observe other people's responses.
  E.   **Dramaturgical analysis** is the study of social interaction that compares everyday life to a theatrical presentation.
       1.   This perspective was initiated by Erving Goffman, who suggested that day-to-day interactions have much in common with being on stage or in a dramatic production.
       2.   Most of us engage in impression management, or **presentation of self** -- people's efforts to present themselves to others in ways that are most favorable to their own interests or image.
       3.   Social interaction, like a theater, has a front stage -- the area where a player performs a specific role before an audience -- and a back stage -- the area where a player is not required to perform a specific role because it is out of view of a given audience.

84

F. The Sociology of Emotions
  1. Arlie Hochschild suggests that we acquire a set of feeling rules, which shape the appropriate emotions for a given role or specific situation.
  2. Emotional labor occurs when employees are required by their employers to feel and display only certain carefully selected emotions.
  3. Gender, class, and race are related to the expression of emotions necessary to manage one's feelings.
G. **Nonverbal communication** is the transfer of information between persons without the use of speech (e.g. facial expressions, head movements, body positions, and other gestures).
H. **Personal space** is the immediate area surrounding a person that the person claims as private. Age, gender, kind of relationship, and social class are important factors in allocation of personal space. Power differentials between people are reflected in personal space and privacy.

V. CHANGING SOCIAL STRUCTURE AND INTERACTION IN THE TWENTY-FIRST CENTURY
  A. The social structure in the U.S. has been changing rapidly in the past decades (e.g., more possible statuses for persons to occupy and roles to play than at any other time in history).
  B. Ironically, at a time when we have more technological capability, more leisure activities and types of entertainment, and vast quantities of material goods available for consumption, many people experience high levels of stress, fear for their lives because of crime, and face problems such as homelessness.
  C. While some individuals and groups continue to show initiative in trying to solve some of our pressing problems, the future of this country rests on our collective ability to deal with major social problems at both the macrolevel (structural) and the microlevel of society.

ANALYZING AND UNDERSTANDING THE BOXES

After reading the chapter and studying the outline, re-read the four boxes and write down key points and possible questions for class discussion.

Sociology and Everyday Life -- "How Much Do You Know About Homeless Persons?"

Key Points:

Discussion Questions:

1.

2.

3.

Sociology and Law -- "Homeless Rights Versus Public Space"

Key Points:

Discussion Questions:

1.

2.

3.

Sociology and Media -- "The Homeless and the Holidays"

Key Points:

Discussion Questions:

1.

2.

3.

Sociology in Global Perspective -- "Homelessness in Japan and France"

Key Points:

Discussion Questions:

1.

2.

3.

PRACTICE TEST

MULTIPLE CHOICE QUESTIONS

Select the response that best answers the question or completes the statement:

1.   The stable patterns of social relationships that exist within a particular
        group or society are referred to as: (p. 160)
        a.   social structure.
        b.   social interaction.
        c.   social dynamics.
        d.   social constructions of reality.

2.   According to the text, homelessness is an example of: (p. 162)
        a.   social depletion.
        b.   social marginality.
        c.   social disintegration.
        d.   social misappropriation.

3.   A _____ is a socially defined position in a group or society. (p. 163)
        a.   location
        b.   set
        c.   status
        d.   role

4.   Maxine is an attorney, a mother, a resident of California, and a Jewish
        American. All of these socially defined positions constitute her:
        (p. 164)
        a.   role pattern.
        b.   role set.
        c.   ascribed statuses.
        d.   status set.

5.   A master status: (p. 164)
     a.   historically has been held only by men.
     b.   is comprised of all of the statuses that a person occupies at a
          given time.
     c.   is the most important status a person occupies.
     d.   is a social position a person always assumes voluntarily.

6.   According to the text, a wedding ring and a Rolls-Royce automobile
     are both examples of: (165)
     a.   conspicuous consumption.
     b.   status symbols.
     c.   status markers.
     d.   master status indicators.

7.   _____ is (are) the dynamic aspect of a status. (p. 166)
     a.   Role
     b.   Norms
     c.   Groups
     d.   People

8.   In her study of women athletes in college sports programs, sociologist
     Tracey Watson found role _____ in the traditionally incongruent
     identities of being a woman and being an athlete. (p. 167)
     a.   strain
     b.   distancing
     c.   conflict
     d.   confusion

9.   According to the text, lesbians and gay men often experience role
     _____ because of the pressures associated with having an
     identity heavily stigmatized by the dominant cultural group. (p. 168)
     a.   strain
     b.   distancing
     c.   conflict
     d.   confusion

10. When a homeless person is able to become a domiciled person,
     sociologists refer to the process as: (p. 169)
     a.   role disengagement.
     b.   role exit.
     c.   role engulfment.
     d.   role relinquishment.

11.　All of the following are examples of a primary group, <u>except</u>: (p. 169)
　　　a.　family.
　　　b.　close friends.
　　　c.　peer groups.
　　　d.　students in a lecture hall.

12.　Which of the following statements <u>best</u> describes the characteristics of a secondary group? (p. 169)
　　　a.　a small, less specialized group in which members engage in impersonal, goal-oriented relationships over an extended period of time.
　　　b.　a small, less specialized group in which members engage in face-to-face, emotion-based interactions over an extended period of time.
　　　c.　a larger, more specialized group in which members engage in face-to-face, emotion-based interactions over an extended period of time.
　　　d.　a larger, more specialized group in which members engage in impersonal, goal-oriented relationships over an extended period of time.

13.　According to the text, the Salvation Army and other caregiver groups that provide services for the homeless and others in need are examples of: (p. 170)
　　　a.　primary groups.
　　　b.　formal organizations.
　　　c.　informal organizations.
　　　d.　social networks.

14.　Sociologists use the term _____ to refer to a set of organized beliefs and rules that establish how a society will attempt to meet its basic social needs. (p. 171)
　　　a.　social structure
　　　b.　social expectations
　　　c.　social networking
　　　d.　social institutions

15.　Emile Durkheim referred to the social cohesion found in industrial societies as: (p. 173)
　　　a.　organic solidarity.
　　　b.　mechanical solidarity.
　　　c.　*Gemeinschaft*.
　　　d.　*Gesellschaft*.

16. The United States is an example of: (p. 174)
   a. *Gemeinschaft.*
   b. *Gesellschaft.*
   c. mechanical solidarity.
   d. postmechanical solidarity.

17. A young person who has been told repeatedly that she or he is not a good student and, as a result, stops studying and receives failing grades is an example of a(n): (p. 177)
   a. selective perception.
   b. objective deduction.
   c. subjective reality.
   d. self-fulfilling prophecy.

18. Ethnomethodologist Harold Garfinkel assigned different activities to his students to see how breaking the unspoken rules of behavior created confusion. His research involved a series of: (p. 179)
   a. shared expectancies.
   b. breaching experiments.
   c. dramaturgical analyses.
   d. impression managements.

19. According to sociologist Arlie Hochschild, people acquire a set of _____, which shape the appropriate emotions for a given role or specific situation. (p. 182)
   a. role expectations
   b. emotional experiences
   c. feeling rules
   d. nonverbal cues

20. According to anthropologist Edward Hall, _____ distance is the "distance zone" that marks an extremely formal relationship and makes interpersonal communication nearly impossible. (p. 186)
   a. public
   b. social
   c. personal
   d. intimate

TRUE-FALSE QUESTIONS

T   F   1.   Most homeless persons are on the streets by choice or because they were deinstitutionalized by mental hospitals. (p. 160).

T F 2. Sociologists use the term status to refer only to high-level positions in society. (p. 163)

T F 3. Historically, the most common master statuses for women have related to positions in the family, such as daughter, wife, and mother. (p. 164)

T F 4. Role performance does not always match role expectation. (p. 166)

T F 5. Role conflict occurs when incompatible role demands are placed on a person by two or more statuses held at the same time. (p. 167)

T F 6. Social networks work the same way for men and women and for people from different racial-ethnic groups. (p. 170)

T F 7. Interactionist theorists argue that social institutions maintain the privileges of the wealthy and powerful while contributing to the powerlessness of others. (p. 172)

T F 8. Relationships in *gemeinschaft* societies are based on achieved rather than ascribed status. (p. 173)

T F 9. Race/ethnicity, gender, and social class play a part in the meanings we give to our interactions with others. (p. 176)

T F 10. The need for impression management is most intense when role players have widely divergent or devalued statuses. (p. 181)

SOCIOLOGY IN OUR TIMES: DIVERSITY ISSUES

1. According to the text, "people of color are overrepresented among the homeless." (p. 160) Using your sociological imagination, how is this "personal trouble" linked to other social problems in our society? If you were to find yourself temporarily homeless, how would you resolve your problem? Would you expect help from a social agency or governmental bureaucracy? Why or why not?

2. The text states that "being poor or rich is a master status that influences many other areas of life, including health, education, and

life opportunities." (p. 164) From your own experience, has your family's social class affected your master status? Do you think your social class has affected your health, education, and life opportunities?

3. Have you experienced role conflict? Role strain? Were your experiences related to your gender, race/ethnicity, class, or age?

4. According to the text, people at the middle- and upper-class levels tap social networks to find employment, to make business deals, and to win political elections. (p. 170) Do you think it is possible for people from working-class and low-income backgrounds to develop effective social networks? Why or why not?

CHAPTER FIVE CROSSWORD PUZZLE

For those who enjoy crossword puzzles, here is a puzzle that contains words and names from Chapter Five. Working the puzzle will help you in reviewing the chapter. The answers appear on page 96.

ACROSS

1. Small, less specialized group in which members engage in face-to-face interactions
4. Sociologist who referred to certain behavior as civil inattention
7. _____ fulfilling prophecy
9. A set of behavioral expectations associated with a given status
11. Role _____ occurs when incompatible demands are built into a single status that a person occupies
13. A large, urban society in which social bonds are based on impersonal and specialized relationships
15. Any physical or social attribute or sign that so devalues a person's social identity that it disqualifies that person from full social acceptance
16. _____ communication: transfer of information between persons without the use of speech
17. Functional theorists state that one of the purposes of social institutions is teaching _____ members
21. One of the ascribed statuses listed in text
22. Social marginality is the state of _____ insider and part outsider in the social structure
24. _____ network: series of social relationships that link an individual to others
25. Social structure: the stable patterns of social relationships that _____ within a particular group or society
26. A stigma so _____ a person's social identity that it disqualifies that person from full social acceptance
27. Along with 33 across, he suggested that the survival strategies of unattached, homeless men are the product of their resourcefulness and ingenuity
32. Deference behavior is important in regard to facial expression, _____ contact, and touching
33. See 27 above
34. Social _____: a set of organized beliefs and rules

DOWN

1. The immediate area surrounding a person that the person claims as private
2. Helen _____ Fuchs Ebaugh studied role exit. The second word in her name makes one of these
3. Sociologist who coined the term ethnomethodology
5. Type of solidarity found in industrial societies
6. Sociologist who suggested that we acquire feeling rules
7. Two or more people who interact frequently and share a common identity and a feeling of interdependence
8. Social marginality the state of being part insider and part outsider in the social _____
10. When Charles is really involved in role distancing, he tells his friends (about the restaurant) that he wouldn't want to _____ there
12. When engaged in role performance, _____ a role
14. Number of "distance zones" observed by Edward Hall
18. __ each have both ascribed and achieved statuses
19. Anthropologist who analyzed the physical distance between people speaking to one another: him and members of his family
20. Role _____ occurs when people disengage from social roles that have been central to their self-identity
22. There are four of these in each chapter. Sociology and Law is an example
23. First six letters of Tonnies' name for a traditional society in which social relationships are based on personal bonds of friendship, etc.
24. According to Goffman, social interaction has a front one and a back one
28. Status _____: all of the statuses that a person occupies at a given time
29. In breaching experiments, ethnomethodologists act like they do ____ understand some basic rule of social life
30. A role ____ of behavioral expectations associated with a given status
31. Role expectation: how a role ought _____ be played

93

ANSWERS TO PRACTICE TEST, CHAPTER 5

Answers to Multiple Choice Questions

1.   a   The stable patterns of social relationships that exist within a particular group or society are referred to as social structure. (p. 160)

2.   b   According to the text, homelessness is an example of social marginality. (p. 162)

3.   c   A status is a socially defined position in a group or society. (p. 163)

4.   d   Maxine is an attorney, a mother, a resident of California, and a Jewish American. All of these socially defined positions constitute her status set. (p. 164)

5.   c   A master status is the most important status a person occupies. (p. 164)

6.   b   According to the text, a wedding ring and a Rolls-Royce automobile are both examples of status symbols. (p. 165)

7.   a   Role is the dynamic aspect of a status. (p. 166)

8.   c   In her study of women athletes in college sports programs, sociologist Tracey Watson found role conflict in the traditionally incongruent identities of being a woman and being an athlete. (p. 167)

9.   a   According to the text, lesbians and gay men often experience role strain because of the pressures associated with having an identity heavily stigmatized by the dominant cultural group. (p. 168)

10.   b   When a homeless person is able to become a domiciled person, sociologists refer to the process as role exit. (p. 169)

11.   d   All of the following are examples of a primary group, <u>except</u>: students in a lecture hall. (p. 169)

12.   d   The statement that <u>best</u> describes the characteristics of a secondary group is: a larger, more specialized group in which members engage in impersonal, goal-oriented relationships over an extended period of time. (p. 169)

13.   b   According to the text, the Salvation Army and other caregiver groups that provide services for the homeless and others in need are examples of formal organizations. (p. 170)

14.   d   Sociologists use the term social institutions to refer to a set of organized beliefs and rules that establish how a society will attempt to meet its basic social needs. (p. 171)

15.   a   Emile Durkheim referred to the social cohesion found in industrial societies as organic solidarity. (p. 173)

16.   b   The United States is an example of *gesellschaft*. (p. 174)

17.   d   A young person who has been told repeatedly that she or he is not a good student and, as a result, stops studying and

receives failing grades is an example of a self-fulfilling prophecy. (p. 177)

18.   b   Ethnomethodologist Harold Garfinkel assigned different activities to his students to see how breaking the unspoken rules of behavior created confusion. His research involved a series of breaching experiments. (p. 179)

19.   c   According to sociologist Arlie Hochschild, people acquire a set of feeling rules , which shape the appropriate emotions for a given role or specific situation. (p. 182)

20.   a   According to anthropologist Edward Hall, public distance is the "distance zone" that marks an extremely formal relationship and makes interpersonal communication nearly impossible. (p. 186)

Answers to True-False Questions

1.   False -- Most homeless persons are not on the streets by choice or because of deinstitutionalization; many hold full- or part-time jobs but earn too little to find an affordable place to live. (p. 160)

2.   False -- In a sociological sense, the term status refers to a socially defined position (regardless of whether it is high-level, mid-level, or low-level) in a group or society characterized by certain expectations, rights, and duties. (p. 163)

3.   True (p. 164)
4.   True (p. 166)
5.   True (p. 167)

6.   False -- According to sociologists, social networks work differently for women and men, for different races/ethnicities, and for members of different social classes because of exclusion from "old-boy" social networks. (p. 170)

7.   False -- Conflict theorists argue that social institutions maintain the privileges of the wealthy and powerful while contributing to the powerlessness of others. (p. 172)

8.   False -- Relationships in Gemeinschaft societies are based on ascribed rather than achieved status. (p. 173)

9.   True (p. 176)
10.   True (p. 181)

ANSWER TO CHAPTER FIVE CROSSWORD PUZZLE

```
P R I M A R Y G R O U P . G O F F M A N . S E L F .
E . . O . A . . . . . . . R . . . R . . . O . . . S
R O L E . S T R A I N . . G E S E L L S C H A F T
S . . A . E . . F A . N O N V E R B A L . . H . R
N E W . . S T I G M A . F O . . . . . . . . . U
A . E . . . K . L . . R A C E . B E I N G P A R T
L . X . . E A . . G . . . . . O . . . R . L . C
. S O C I A L . . E X I S T . X . . R O . L . U
P . T . . . . . M . . . . D E V A L U E S . . R
A . T A . I T . A N D E R S O N S . . . S . P E
C . G I . . . G I E . . . O . . . . . .
E Y E . S N O W . I N S T I T U T I O N
```

96

CHAPTER 6
GROUPS AND ORGANIZATIONS

BRIEF CHAPTER OUTLINE

CHAPTER SUMMARY

Groups are a key element of our social structure and much of our social interaction takes place within them. A **social group** is a collection of two or more people who interact frequently, share a sense of belonging, and have a feeling of interdependence. Social groups may be either **primary groups** -- small, personal groups in which members engage in emotion-based interactions over an extended period -- or **secondary groups** -- larger, more specialized groups in which members have less personal and more formal, goal-oriented relationships. All groups set boundaries to indicate who does and who does not belong: an **ingroup** is a group to which we belong and with which we identify; an **outgroup** is a group we do not belong to or perhaps feel hostile toward. The size of a group is one of its most important features. The smallest groups are **dyads** -- groups composed of two members -- and **triads** -- groups of three. In order to maintain ties with a group, many members are willing to conform to norms established and reinforced by group members. Three types of **formal organizations** -- highly structured secondary groups formed to achieve specific goals in an efficient manner -- are normative, coercive, and utilitarian organizations. A **bureaucracy** is a formal organization characterized by hierarchical authority, division of labor, explicit procedures, and impersonality in personnel concerns. The **iron law of oligarchy** refers to the tendency of organizations to become a bureaucracy ruled by the few. A recent movement to humanize bureaucracy has focused on developing human resources. As we approach the twenty-first century, all of us benefit from organizations that operate humanely and that include opportunities for all, regardless of race, gender, or class.

LEARNING OBJECTIVES

After reading Chapter 6, you should be able to:

1. Distinguish between aggregates, categories, and groups from a sociological perspective.

2. Distinguish between primary and secondary groups and explain how people's relationships differ in each.

3. State definitions for ingroup, outgroup, and reference group and describe the significance of these concepts in everyday life.

4. Contrast functionalist and conflict perspectives on the purposes of groups.

5. Describe dyads and triads and explain how interaction patterns change as the size of a group increases.

6. Distinguish between the two functions of leadership and the three major styles of group leadership.

7. Describe the experiments of Solomon Asch and Stanley Milgram and explain their contributions to our understanding about group conformity and obedience to authority.

8. Explain what is meant by groupthink and discuss reasons why it can be dangerous for organizations.

9.   Compare normative, coercive, and utilitarian organizations and describe the nature of membership in each.

10.   Summarize Max Weber's perspective on rationality and outline his ideal characteristics of bureaucracy.

11.   Describe the informal structure in bureaucracies and list its positive and negative aspects.

12.   Discuss the major shortcomings of bureaucracies and their effects on workers, clients or customers, and levels of productivity.

13.   Describe the iron law of oligarchy and explain why bureaucratic hierarchies and oligarchies go hand in hand.

14.   Evaluate U.S. and Japanese models of organization.

KEY TERMS (defined at page number shown and in glossary)

aggregate 195	ideal type 212
authoritarian leaders 201	informal structure 213
bureaucracy 211	ingroup 196
bureaucratic personality 216	instrumental leadership 200
category 195	iron law of oligarchy 222
conformity 201	laissez-faire leaders 201
democratic leaders 201	outgroup 196
dyad 199	rationality 211
expressive leaders 201	reference group 198
goal displacement 216	small group 199
groupthink 206	triad 199

CHAPTER OUTLINE

I.   SOCIAL GROUPS

    A.   Groups, Aggregates, and Categories

        1.   A **social group** is a collection of two or more people who interact frequently with one another, share a sense of belonging, and have a feeling of interdependence.

        2.   An **aggregate** is a collection of people who happen to be in the same place at the same time but share little else in common.

        3.   A **category** is a number of people who may never have met one another but share a similar characteristic.

    B.   Types of Groups

        1.   Primary and Secondary Group

            a.   According to Charles H. Cooley, a **primary** group is a small group whose members engage in face-to-face, emotion-based interactions over an extended period of time.

            b.   A **secondary group** is a larger, more specialized group in which the members engage in more impersonal, goal-oriented relationships for a limited period of time.

        2.   Ingroups and Outgroups

            a.   According to William Graham Sumner, an **ingroup** is a group to which a person belongs and with which the person feels a sense of identity.

            b.   An **outgroup** is a group to which a person does not belong and toward which the person may feel a sense of competitiveness or hostility.

        3.   Reference Groups

            a.   A **reference group** is a group that strongly influences a person's behavior and social attitudes, regardless of whether that individual is an actual member.

        b.   Reference groups help us explain why our behavior and attitudes sometimes differ from those of our membership groups; we may accept the values and norms of a group with which we identify rather than one to which we belong.

II.   GROUP CHARACTERISTICS AND DYNAMICS

    A.   Group Size

        1.   A **small group** is a collectively small enough for all members to be acquainted with one another and to interact simultaneously.

        2.   According to Georg Simmel, small groups have distinctive interaction patterns which do not exist in larger groups.

            a.   In a **dyad** -- a group composed of two members -- the active participation of both members is crucial for the group's survival and members have a more intense bond and a sense of unity not found in most larger groups.

            b.   When a third person is added to a dyad, a **triad** -- a group composed of three members -- is formed, and the nature of the relationship and interaction patterns change.

        3.   As group size increases, members tend to specialize in different tasks, and communication patterns change.

    B.   Group Leadership

        1.   Leaders are responsible for directing plans and activities so that the group completes its task or fulfills its goals.

        2.   Leadership functions:

            a.   **Instrumental leadership** is goal or task oriented; if the underlying purpose of a group is to complete a task or reach a particular goal, this type of leadership is most appropriate.

            b.   **Expressive leadership** provides emotional support for members; this type of leadership is most appropriate when harmony, solidarity, and high morale are needed.

        3.   Leadership styles:

            a.   **Authoritian leaders** make all major group decisions and assign tasks to members.

            b.   **Democratic leaders** encourage group discussion and decision-making through consensus building.

     c. **Laissez-faire leaders** are only minimally involved in decision-making and encourage group members to make their own decisions.

  C. Group Conformity

    1. **Conformity** is the process of maintaining or changing behavior to comply with the norms established by a society, subculture, or other group.

    2. In a series of experiments, Solomon Asch found that the pressure toward group conformity was so great that participants were willing to contradict their own best judgment if the rest of the group disagreed with them.

    3. Stanley Milgram (a former student of Asch's) conducted a series of controversial experiments and concluded that people's obedience to authority may be more common than most of us would like to believe.

    4. Irving Janis coined the term **groupthink** to describe the process by which members of a cohesive group arrive at a decision that many individual members privately believe is unwise.

III. FORMAL ORGANIZATIONS

  A. A **formal organization** is a highly structured secondary group formed for the purpose of achieving specific goals in the most efficient manner (e.g., corporations, schools, and government agencies).

  B. Types of Formal Organizations

    1. Amitai Etzioni classified formal organizations into three categories based on the nature of membership.

    2. We voluntarily join **normative organizations** when we want to pursue some common interest or to gain personal satisfaction or prestige from being a member. Examples include political parties, religious organizations, and college social clubs.

    3. People do not voluntarily become members of **coercive organizations** -- associations people are forced to join. Examples include total institutions, such as boot camps and prisons.

    4. We voluntarily join **utilitarian organizations** when they can provide us with a material reward we seek. Examples include colleges and universities, and the workplace.

  C. Bureaucracies

    1. **Bureaucracy** is an organizational model characterized by a hierarchy of authority, a clear

division of labor, explicit rules and procedures, and impersonality in personnel matters.

2.   According to **Max Weber**, bureaucracy is the most "rational" and efficient means of attaining organizational goals because it contributes to coordination and control. **Rationality** is the process by which traditional methods of social organization, characterized by informality and spontaneity, gradually are replaced by efficiently administered formal rules and procedures.

3.   An **ideal type** is an abstract model which describes the recurring characteristics of some phenomenon; the ideal characteristics of bureaucracy include:

a.   Division of Labor-- each member has a specific status with certain assigned tasks to fulfill.

b.   Hierarchy of Authority -- a chain of command that is based on each lower office being under the control and supervision of a higher one.

c.   Rules and Regulations -- standardized rules and regulations establish authority within an organization and usually are provided to members in a written format.

d.   Employment Based on Technical Qualifications -- hiring of staff members and professional employees is based on specific qualifications; individual performance is evaluated against specific standards; and promotions are based on merit as spelled out in personnel policies.

e.   Impersonality -- interaction is based on status and standardized criteria rather than personal feelings or subjective factors.

4.   An organization's **informal structure** is composed of those aspects of participants' day-to-day activities and interactions that ignore, bypass, or do not correspond with the official rules and procedures of the bureaucracy.

a.   The Hawthorne studies made social scientists aware of the presence of informal networks and their effect on workers' productivity; they concluded that the level of productivity was determined by the workers' informal networks, not by the levels set by management.

        b.   There are two schools of thought about informal structure in organizations; one emphasizes control (or eradication) of informal groups; the other suggests that they should be nurtured.

    D.   Shortcomings of Bureaucracy

        1.   Inefficiency and Rigidity

            a.   **Goal displacement** occurs when the rules become an end-in-themselves (rather than a means-to-an-end), and organizational survival becomes more important than achievement of goals

            b.   The term **bureaucratic personality** is used to describe those workers who are more concerned with following correct procedures than they are with getting the job done correctly.

        2.   Resistance to Change

        3.   Perpetuation of Race, Class, and Gender Inequalities

    E.   Bureaucracy and Oligarchy

        1.   Max Weber believed that bureaucracy was a necessary evil because it achieved coordination and control and thus efficiency in administration; however, he believed such organizations stifled human initiative and creativity.

        2.   Bureaucracy generates an enormous degree of unregulated and often unperceived social power in the hands of a very few leaders.

        3.   According Robert Michels, this results in the **iron law of oligarchy** -- a bureaucracy ruled by a few people.

IV.   AN ALTERNATIVE FORM OF ORGANIZATION

    A.   "Humanizing" bureaucracy includes: (1) less rigid, hierarchical structures and greater sharing of power and responsibility; (2) encouragement of participants to share their ideas and try new approaches; and (3) efforts to reduce the number of people in dead-end jobs and to help people meet outside family responsibilities while still receiving equal treatment inside the organization.

    B.   The Japanese model of organization has been widely praised for its innovative structure, which (until recently) has included:

        1.   Lifetime Employment-- Workers were guaranteed permanent employment after an initial probationary period.

2. Quality Circles -- small workgroups that meet regularly with managers to discuss the group's performance and working conditions.

V. NEW ORGANIZATIONS FOR THE TWENTY-FIRST CENTURY?

A. There is a lack of consensus among organizational theorists about the "best" model of organization; however, some have suggested a horizontal model in which both hierarchy and functional or departmental boundaries largely would be eliminated.

B. In the horizontal structure, a limited number of senior executives would still exist in support roles (such as finance and human resources); everyone else would work in multidisciplinary teams that would perform core processes (e.g., product development or sales).

C. It is difficult to determine what the best organizational structure for the future might be; however, everyone can benefit from humane organizational environments that provide opportunities for all people regardless of race, gender, or class.

ANALYZING AND UNDERSTANDING THE BOXES

After reading the chapter and studying the outline, re-read the four boxes and write down key points and possible questions for class discussion.

Sociology and Law -- "Sexual Harassment and Congress"

Key Points:

Discussion Questions:

1.

2.

3.

Sociology and Everyday Life -- "How Much Do You Know About Sexual Harassment?"

Key Points:

Discussion Questions:

1.

2.

3.

Sociology and Media -- "The 5 Percent Factor: Sexual Harassment, Media Style"

Key Points:

Discussion Questions:

1.

2.

3.

Sociology in Global Perspective -- "Sexual Harassment Crosses Global Boundaries"

Key Points:

Discussion Questions:

1.

2.

3.

PRACTICE TEST

MULTIPLE CHOICE QUESTIONS

Select the response that best answers the question or completes the statement:

1.   A woman, who has been a junior executive for several years and is now eager for promotion, receives sexual advances from the vice president of the corporation. She is told that her promotion will be forthcoming if she agrees to have sex with the V.P. This situation exemplifies _____ harassment. (p. 194)
    a.   informal sexual
    b.   arbitrary sexual
    c.   quid pro quo
    d.   hostile environment

2.   A(n) _____ is a collection of people who happen to be in the same place at the same time while a(n) _____ is a number of people who may never have met one another but share a similar characteristic. (p. 195)
    a.   aggregate - category
    b.   category - aggregate
    c.   social group - aggregate
    d.   category - social group

3.   John had thought about trying out for the football team, but because he is not very athletic, he decides to hang out with the "stoners" at his school. He harbors some feelings of hostility for the football team. For John, the football team represents a(n): (p. 196)
    a.   reference group.
    b.   secondary group.
    c.   ingroup.
    d.   outgroup.

4.   A _____ is an alliance created in an attempt to reach a shared objective or goal. (p. 199)
    a.   triad
    b.   coalition
    c.   dyad
    d.   reference group

5. _____ leadership is goal or task oriented; _____ leadership provides emotional support for members. (pp. 200-201)
    a. authoritarian - democratic
    b. authoritarian - laissez-faire
    c. expressive - instrumental
    d. instrumental - expressive

6. A group leader at a business-related seminar is only minimally involved with the decisions made by the group and encourages members to make their own choices. This illustrates a(n) _____ leader. (p. 201)
    a. expressive
    b. democratic
    c. laissez-faire
    d. authoritarian

7. In one of Solomon Asch's experiments, a subject was asked to compare the length of lines printed on a series of cards without knowing that all other research "subjects" actually were assistants to the researcher. The results of Asch's experiments revealed that: (pp. 201-202)
    a. 85 percent of the subjects routinely chose the correct response regardless of the assistants' opinions.
    b. 33 percent of the subjects routinely chose to conform to the opinion of the assistants by giving an incorrect answer.
    c. 10 percent of the subjects routinely chose to conform to the opinion of the assistants by giving an incorrect answer.
    d. the opinion of the assistants had no influence on the subjects' opinions.

8. In reference to the experiments by Stanley Milgram, the text points out that: (pp. 203-204)
    a. obedience to authority may be less common than most people would like to believe.
    b. many of the subjects questioned the ethics of the experiment.
    c. research such as this raises important questions concerning research ethics.
    d. most of the subjects were afraid to conform because the use of electrical current was involved.

9.   According to the text, the "Bay of Pigs" fiasco during the Kennedy presidential administration is an example of: (p. 207)
     a.   groupthink.
     b.   obedience to authority.
     c.   compliance.
     d.   the iron law of oligarchy.

10. All of the following are types of formal organizations, <u>except</u> _____ organizations. (p. 209)
     a.   normative
     b.   anomic
     c.   coercive
     d.   utilitarian

11. Membership in _____ organizations is involuntary. (p. 210)
     a.   normative
     b.   anomic
     c.   coercive
     d.   utilitarian

12. According to Max Weber, rationality refers to: (p. 211)
     a.   the process by which bureaucracy is gradually replaced by alternative types of organization such as quality circles.
     b.   the process by which traditional methods of social organization are gradually replaced by bureaucracy.
     c.   the level of sanity (or insanity) of people in an organization.
     d.   the logic used by organizational leaders in decision making.

13. All of the following are ideal-type characteristics of bureaucratic organizations, as specified by Max Weber, <u>except</u>: (pp. 212-213)
     a.   coercive leadership.
     b.   impersonality.
     c.   hierarchy of authority.
     d.   division of labor.

14. In higher education, student handbooks, catalogs, and course syllabi are examples of: (p. 213)
     a.   informal structures in bureaucracy.
     b.   impersonality.
     c.   hierarchy of authority.
     d.   rules and regulations.

15. In the Hawthorne studies, workers in the "bank wiring room": (p. 214)
    a. worked harder when management offered them financial incentives.
    b. adhered to the formal rules of the organization because they believed they would lose their jobs otherwise.
    c. tended to work slowly in the morning but work rapidly in the afternoon when managers were most likely to see them.
    d. determined the level of productivity within informal networks, not by the levels set by management.

16. The _____ approach to management views informal networks as a type of adaptive behavior workers engage in because they experience a lack of congruence between their own needs and the demands of the organization. (p. 215)
    a. authoritarian
    b. human relations
    c. public relations
    d. quality circles

17. All of the following are shortcomings of bureaucracy, except: (p. 216)
    a. impersonality
    b. inefficiency and rigidity
    c. resistance to change
    d. perpetuation of race, class, and gender inequalities

18. The statement that best expresses the Peter Principle is: (pp. 218-219)
    a. People rise to the level of their incompetence.
    b. Work expands to fill the time available for its completion.
    c. When things can go wrong, they will.
    d. If it ain't broke, don't fix it.

19. According to sociologist Joe R. Feagin's recent research on racial and ethnic inequalities in organizations, many middle-class African Americans today: (p. 219)
    a. are included in informal communications networks at work.
    b. have mentors who take an interest in furthering their careers.
    c. have found that entry into dominant white bureaucratic organizations has brought about integration for them.
    d. have experienced an internal conflict between the ideal of equal opportunity and the prevailing norms of many organizations.

20. According to Robert Michels, all organizations encounter: (p. 222)
   a. the Peter Principle.
   b. the iron law of oligarchy.
   c. Murphy's Law.
   d. groupthink.

TRUE-FALSE QUESTIONS

T   F   1.   Sexual harassment always occurs in a one-on-one setting. (p. 194)

T   F   2.   Although formal organizations are secondary groups, they also contain many primary groups within them. (p. 196)

T   F   3.   According to functionalists, people form groups to meet instrumental and expressive needs. (p. 198)

T   F   4.   If one member withdraws from a dyad, the group ceases to exist. (p. 199)

T   F   5.   Laissez-faire leaders encourage group discussion and decision making through consensus building. (p. 201)

T   F   6.   Recent research on sexual harassment by psychologist John Pryor suggests that there is a relationship between group conformity and harassment. (pp. 204-205)

T   F   7.   Political parties, parent-teacher associations, and college sororities and fraternities are examples of utilitarian organizations. (p. 209)

T   F   8.   When many people think of bureaucracy, they think of a high level of productivity and efficiency. (p. 211)

T   F   9.   An ideal type is an abstract model that describes the recurring characteristics of some phenomenon. (p. 212)

T   F   10.   In a university, division of labor between the faculty and administration often is very blurred. (p. 212)

T F 11. Student-faculty relationships are based on both hierarchical and professional authority patterns. (p. 213)

T F 12. Standardized test scores for admission to colleges and universities are an example of bureaucratic impersonality. (p. 213)

T F 13. The formal structure of an organization has been referred to as "bureaucracy's other face." (p. 213)

T F 14. The Hawthorne studies made sociologists aware that research subjects modify their behavior when they know they are being observed and that informal networks have an effect on workers' productivity. (p. 214)

T F 15. Traditional management theories are based on the assumption that people basically are lazy and motivated by greed. (p. 215)

SOCIOLOGY IN OUR TIMES: DIVERSITY ISSUES

1. The text states that women and men often view sexual harassment from entirely different perspectives. (p. 194) Do you think this statement applies to you and your friends, co-workers, professors, bosses, and other acquaintances? Why or why not?

2. Can you think of organizations on your campus or in your community where criteria for membership include the ability to pay initiation fees and/or membership dues? (p. 197) To what extent is class a determining factor in gaining membership in these groups? Do gender, race/ethnicity, religion, age, and other factors figure into the membership equation?

3. If you are a student of color, analyze your experiences in organizations to see if you agree with the statement in your text (p. 219) that "some bureaucracies perpetuate inequalities of race, class, and gender because this form of organizational structure creates a specific type of work or learning environment." Can you think of a situation where you found the organizational "climate" to be "chilly" because of your race/ethnicity?

If you are a white student, analyze your experiences in organizations to see if you can think of a situation in which a person of color was treated differently (or "indifferently") because of her/his race/ethnicity.

112

4.   If you are a student from a low-income family, analyze your experiences in organizations in regard to the statement quoted above. Likewise, if you are a student from a middle- to upper-income family, see if you can think of a situation in which a low-income person was treated differently from you because of lack of economic means.

5.   If you are a woman, analyze your experiences in organizations in regard to the statement quoted above. Likewise, if you are a man, see if you can think of a situation in which a woman was treated differently from you because she was a woman.

CHAPTER SIX CROSSWORD PUZZLE

For those who enjoy crossword puzzles, here is a puzzle that contains words and names from Chapter Six. Working the puzzle will help you in reviewing the chapter. The answers appear on page 117.

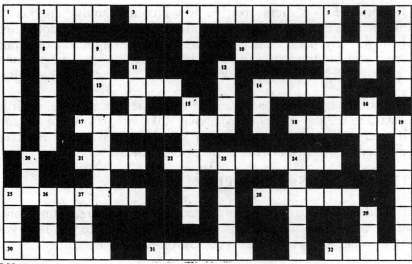

ACROSS

1. He used the term "primary group" to describe a small, less specialized group
3. Type of leadership that is goal or task oriented
8. A group composed of three members
10. Authoritarian leaders demand _____ from others
13. First name of sociologist who suggested that small groups have interaction patterns that do not exist in larger groups
14. He coined the term "groupthink"
17. A _____ group is one that influences a person's behavior and attitudes even if not a member
18. He classified social groups as normative, coercive, and utilitarian
21. Instrumental leadership is goal or _____ oriented
22. Process by which members of a cohesive group may arrive at some decisions
25. Bureaucratic alienation: workers at this point lose ____ in what is happening around them
28. She analyzed how the power structure of bureaucracies can negatively impact subordinate groups
30. As the ____ of people in a group increases, it becomes increasingly difficult for everyone to participate in the same conversation
31. He suggested that small groups have distinctive interaction patterns.
32. For several decades, the Japanese _____ of organization has been highly praised for its innovative structure

DOWN

1. A number of people who may never have met one another but share a similar characteristic
2. A group to which a person does not belong and toward which the person may feel competitive or hostile
4. A dyad has this many members
5. _____-faire leaders are only minimally involved in decision making
6. _____ individual members of a group may think a groupthink decision is unwise
7. _____ type: an abstract model that describes the recurring characteristics of some phenomenon
9. A collection of people who happen to be at the same place at the same time
11. A secondary group is a larger, _____ specialized group
12. In Asch's research, subjects were asked to indicate which _____ on the second card was identical in length to the one on the first card
14. A bureaucratic personality is more concerned with correct procedures than with getting the _____ done
15. Group to which a person belongs and feels a sense of identity
16. _____ displacement occurs when the rules become an end in themselves instead of a [answer to 29 Down]
19. An organization's _____ structure is composed of activities that may ignore rules and procedures
20. Authoritarian leaders "_____ the show"
23. Members of the group in 22 across may think some decisions are _____
24. Participants may _____ rules of bureaucracy when participating in an organization's informal structure
25. The _____ law of oligarchy
26. An ingroup is "us"; and outgroup is "_____"
27. Weber asserted that _____s and regulations establish authority within an organization
29. Means to an _____

114

ANSWERS TO PRACTICE TEST, CHAPTER 6

Answers to Multiple Choice Questions

1.   c   A woman, who has been a junior executive for several years and is now eager for promotion, receives sexual advances from the vice president of the corporation. She is told that her promotion will be forthcoming if she agrees to have sex with the V.P. This situation exemplifies quid pro quo harassment. (p. 194)

2.   a   An aggregate is a collection of people who happen to be in the same place at the same time while a category is a number of people who may never have met one another but share a similar characteristic. (p. 195)

3.   d   John had thought about trying out for the football team, but because he is not very athletic, he decides to hang out with the "stoners" at his school. He harbors some feelings of hostility for the football team. For John, the football team represents an outgroup. (p. 196)

4.   b   A coalition is an alliance created in an attempt to reach a shared objective or goal. (p. 199)

5.   d   Instrumental leadership is goal or task oriented; expressive leadership provides emotional support for members. (pp. 200-201)

6.   c   A group leader at a business-related seminar is only minimally involved with the decisions made by the group and encourages members to make their own choices. This illustrates a laissez-faire leader. (p. 201)

7.   b   In one of Solomon Asch's experiments, a subject was asked to compare the length of lines printed on a series of cards without knowing that all other research "subjects" actually were assistants to the researcher. The results of Asch's experiments revealed that 33 percent of the subjects routinely chose to conform to the opinion of the assistants by giving an incorrect answer. (pp. 201-202)

8.   c   In reference to the experiments by Stanley Milgram, the text points out that research such as this raises important questions concerning research ethics. (pp. 203-204)

9.   a   According to the text, the "Bay of Pigs" fiasco during the Kennedy presidential administration is an example of groupthink. (p. 207)

10.   b   All of the following are types of formal organizations, <u>except</u>: anomic organizations. (p. 209)

11.   c   Membership in coercive organizations is involuntary. (p. 210)

12. b According to Max Weber, rationality refers to the process by which traditional methods of social organization are gradually replaced by bureaucracy. (p. 211)

13. a All of the following are ideal-type characteristics of bureaucratic organizations, as specified by Max Weber, except: coercive leadership. (pp. 212-213)

14. d In higher education, student handbooks, catalogs, and course syllabi are examples of rules and regulations. (p. 213)

15. d In the Hawthorne studies, workers in the "bank wiring room" determined the level of productivity within informal networks, not by the levels set by management. (p. 214)

16. b The human relations approach to management views informal networks as a type of adaptive behavior workers engage in because they experience a lack of congruence between their own needs and the demands of the organization. (p. 215)

17. a All of the following are shortcomings of bureaucracy, except: impersonality. (p. 216)

18. a The statement that best expresses the Peter Principle is: People rise to the level of their incompetence. (pp. 218-219)

19. d According to sociologist Joe R. Feagin's recent research on racial and ethnic inequalities in organizations, many middle-class African Americans today have experienced an internal conflict between the ideal of equal opportunity and the prevailing norms of many organizations. (p. 219)

20. b According to Robert Michels, all organizations encounter the iron law of oligarchy. (p. 222)

Answers to True-False Questions
1. False -- Although sexual harassment may occur in a one-on-one setting, harassment also is found in small and large groups. (p. 194)
2. True (p. 196)
3. True (p. 198)
4. True (p. 199)
5. False -- Democratic leaders encourage group discussion and decision making through consensus building; laissez-faire leaders are only minimally involved in decision making and encourage group members to make their own decisions. (p. 201)
6. True (pp. 204-205)
7. False -- Political parties, parent-teacher associations, and college sororities and fraternities are examples of normative organizations. (p. 209)
8. False -- When many people think of bureaucracy, they think of "buck-passing" and "red tape." (p. 211)
9. True (p. 212)

10. False -- In a university, a distinct division of labor exists between the faculty and administration.
11. True (p. 213)
12. True (p. 213)
13. False -- The <u>informal</u> structure of an organization has been referred to as "bureaucracy's other face." (p. 213)
14. True (p. 214)
15. True (p. 215)

ANSWER TO CHAPTER SIX CROSSWORD PUZZLE

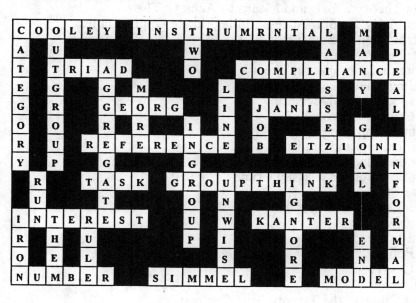

CHAPTER 7
DEVIANCE AND CRIME

BRIEF CHAPTER OUTLINE
What Is Deviance?
Functionalist Perspectives on Deviance
     How is Deviance Functional for Society?
     Strain Theory: Goals and Means to Achieve Them
     Opportunity Theory: Access to Illegitimate Opportunities
     Control Theory: Social Bonding
Interactionist Perspectives on Deviance
     Differential Association Theory
     Labeling Theory
Conflict Perspectives on Deviance
     The Critical Approach
     Feminist Approaches
Crime Classifications and Statistics
     How the Law Classifies Crime
     How Sociologists Classify Crime
     Crime Statistics
     Street Crimes and Criminals
     Crime Victims
The Criminal Justice System
     The Police
     The Courts
     Punishment
Deviance and Crime in the Twenty-First Century

CHAPTER SUMMARY
All societies have norms to reinforce and help teach acceptable behavior. They also have various mechanisms of **social control** --systematic practices developed by social groups to encourage conformity and to discourage **deviance** -- any behavior, belief or condition that violates cultural norms. **Crime** is a form of deviant behavior that violates criminal law and is punishable by fines, jail terms, and other sanctions. Functionalists suggest that deviance is inevitable in all societies and serves several functions: it clarifies rules, unites groups, and promotes social change. Functionalists use **strain theory**, **opportunity theory**, and **social bonding theory** to argue that socialization into the core value of material success without the corresponding legitimate means to achieve that goal accounts for much of the crime committed by people from lower-income backgrounds, especially when a person's ties to society are weakened or broken. Interactionists use

differential association theory and **labeling theory** to explain how a person's behavior is influenced and reinforced by others. Conflict theorists suggest that people with economic and political power define as criminal any behavior that threatens their own interests and are able to use the law to protect their own interests. Various feminist approaches focus on the intertwining of gender, class, race/ethnicity, and deviance. While the law classifies crime into felonies and misdemeanors based on the seriousness of the crime, sociologists categorize crimes according to how they are committed and how society views them. Four general categories of crime include: **conventional** or **street crime**, **occupational** or **white-collar crime**, **organized crime**, and **political crimes**. Studies show that many more crimes are committed than are reported in official crime statistics. Gender, age, class, and race are key factors in official crime statistics. The criminal justice system includes the police, the courts, and prisons; these agencies often have considerable discretion in dealing with deviance. As we move into the twenty-first century, we need new approaches for dealing with crime and delinquency. We also need to insure an equal justice system for all, regardless of race, class, sex, or age.

LEARNING OBJECTIVES
After reading Chapter 7, you should be able to:
1.   Explain the nature of deviance and describe its most common forms.

2.   Discuss the functions of deviance from a functionalist perspective and outline the principal features of strain, opportunity, and control theories.

3.   Describe the key components of differential association theory and labeling theory and explain why they are examples of interactionist perspectives.

4.   Discuss conflict perspectives on deviance and note the strengths and weaknesses of critical and feminist approaches to deviance and crime.

5.   Distinguish between legal and sociological classifications of crime.

6. Describe the types of behavior included in conventional crimes.

7. Differentiate between occupational and corporate crime and explain why people who commit such crimes may not be viewed as "criminals."

8. Describe organized crime and political crime and explain how each may weaken social control in a society.

9. Explain why official crime statistics may not be an accurate reflection of the actual number and kinds of crimes committed in the United States.

10. Describe the criminal justice system and explain how police, courts, and prisons have considerable discretion in dealing with offenders.

11. State the four functions of punishment and explain how disparate treatment of the poor, all people of color, and white women is evident in the U.S. prison system.

KEY TERMS (defined at page number shown and in glossary)
conventional or street crime 247
corporate crime 250
crime 235
criminology 243
deviance 233
differential association
    theory 240
illegitimate opportunity
    structures 238
juvenile delinquency 235
labeling theory 241

medicalization of deviance 263
occupational or white-collar
    crime 249
organized crime 250
political crime 251
primary deviance 241
punishment 261
secondary deviance 241
social bond theory 240
social control 232
strain theory 236

KEY PEOPLE (identified at page number shown)

CHAPTER OUTLINE

I.   WHAT IS DEVIANCE?

    A.   All societies have norms that govern acceptable behavior and mechanisms of **social control** -- systematic practices developed by social groups to encourage conformity and to discourage deviance.

    B.   Deviance is relative and it varies in its degree of seriousness: some forms of deviant behavior are officially defined as a **crime** -- a behavior that violates criminal law and is punishable with fines, jail terms, and other sanctions.

II.   FUNCTIONALIST PERSPECTIVES ON DEVIANCE

    A.   Emile Durkheim regarded deviance as a natural and inevitable part of all societies.

    B.   Deviance is universal because it serves three important functions:

        1.   Deviance clarifies rules.

        2.   Deviance unites a group.

        3.   Deviance promotes social change.

    C.   Functionalists acknowledge that deviance also may be dysfunctional for society; if too many people violate the norms, everyday existence may become unpredictable, chaotic, and even violent.

    D.   According to **strain theory**, people feel strain when they are exposed to cultural goals that they are unable to obtain because they do not have access to culturally approved means of achieving those goals. Robert Merton identified five ways in which people adapt to cultural goals and approved ways of achieving them:

        1.   Conformity;

        2.   Innovation;

        3.   Ritualism;

        4.   Retreatism; and

        5.   Rebellion.

    E.   According to Richard Cloward and Lloyd Ohlin, for deviance to occur people must have access to **illegitimate opportunity**

structures -- circumstances that provide an opportunity for people to acquire through illegitimate activities what they cannot achieve through legitimate channels.

F.   **Social bond theory** holds that the probability of deviant behavior increases when a person's ties to society are weakened or broken.

III.   INTERACTIONIST PERSPECTIVES ON DEVIANCE

A.   **Differential association theory** states that individuals have a greater tendency to deviate from societal norms when they frequently associate with persons who are more favorable toward deviance than conformity.

B.   **Labeling theory** states that deviants are those people who have been successfully labeled as such by others.

         1.   **Primary deviance** is the initial act of rule-breaking.

         2.   **Secondary deviance** occurs when a person who has been labeled a deviant accepts that new identity and continues the deviant behavior.

IV.   CONFLICT PERSPECTIVES ON DEVIANCE

A.   According to conflict theorists, people in positions of power maintain their advantage by using the law to protect their own interests.

B.   According to the critical approach, the way laws are made and enforced benefits the capitalist class by ensuring that individuals at the bottom of the social class structure do not infringe on the property or threaten the safety of those at the top.

C.   While there is no single feminist perspective on deviance and crime, three schools of thought have emerged:

         1.   Liberal feminism is based on the assumption that women's deviance and crime is a rational response to gender discrimination experienced in work, marriage, and interpersonal relationships.

         2.   Radical feminism is based on the assumption that women's deviance and crime is related to patriarchy (male domination over females) that keeps women more tied to family, sexuality, and home, even if women also have full-time paid employment.

         3.   Socialist feminism is based on the assumption that women's deviance and crime is the result of women's exploitation by capitalism and patriarchy (e.g., their overrepresentation in relatively low-wage jobs and their lack of economic resources).

         4.   Feminist scholars of color have pointed out that these schools of feminist thought do not include race and ethnicity in their analyses. As a result, some recent

studies have focused on simultaneous effects of race, class, and gender on the deviant behavior by some women of color.

V.   CRIME CLASSIFICATIONS AND STATISTICS

    A.   Crimes are divided into felonies and misdemeanors based on the seriousness of the crime.

    B.   Sociologists categorize crimes based on how they are committed and how society views the offenses.

        1.   **Conventional** or **street crime** is all violent crime, certain property crimes, and certain morals crimes.

        2.   **Occupational** or **white-collar crime** is illegal activities committed by people in the course of their employment or financial affairs.

        3.   **Corporate crime** is an illegal act committed by corporate employees on behalf of the corporation and with its support.

        4.   **Organized crime** is a business operation that supplies illegal goods and services for profit.

        5.   **Political crime** refers to illegal or unethical acts involving the usurpation of power by government officials, or illegal/unethical acts perpetrated against the government by outsiders seeking to make a political statement, undermine the government, or overthrow it.

    C.   Official crime statistics, such as those found in the *Uniform Crime Report*, provide important information on crime; however, the data reflect only those crimes that have been reported to the police.

        1.   The National Crime Victimization Survey and anonymous self-reports of criminal behavior have made researchers aware that the incidence of some crimes, such as theft, is substantially higher than reported in the UCR.

        2.   Crime statistics do not reflect many crimes committed by persons of upper socioeconomic status in the course of business because they are handled by administrative or quasi-judicial bodies.

    D.   Street Crimes and Criminals

        1.   Gender and Crime

            a.   The three most common arrest categories for both men and women are driving under the influence of alcohol or drugs (DUI), larceny, and minor or criminal mischief types of offenses.

b.   Liquor law violations (such as underage drinking), simple assault, and disorderly conduct are middle range offenses for both men and women, and the rate of arrests for murder, arson, and embezzlement are relatively low for both men and women.

c.   There is a proportionately greater involvement of men in major property crimes and violent crime.

2.   Age and Crime

a.   Arrest rates for index crimes are highest for people between the ages of 13 and 25, with the peak being between ages 16 and 17.

b.   Rates of arrest remain higher for males than females at every age and for nearly all offenses.

3.   Social Class and Crime

a.   Individuals from all social classes commit crimes; they simply commit different kinds of crime.

b.   Persons from lower socioeconomic backgrounds are more likely to be arrested for violent and property crimes; only a very small proportion of individuals who commit white-collar or elite crimes will ever be arrested or convicted.

4.   Race and Crime

a.   In 1993, whites (including Latinos/as) accounted for about 61 percent of all arrests for index crimes; arrest rates for whites were higher in non-violent property crimes such as fraud and larceny-theft, but were lower than the rates for African Americans in violent crimes such as robbery and murder.

b.   In 1993, whites constituted about 65 percent of all arrests for property crimes and almost 52 percent of arrests for violent crimes; African Americans accounted for over 45 percent of arrests for violent crimes and 33 percent of arrests for property crimes.

c.   Arrest records tend to produce over generalizations about who commits crime because arrest statistics are not an accurate reflection of the crimes actually committed in our society.

5. Crime Victims
   a. Men are more likely to victimized by crime although women tend to be more fearful of crime, particularly those directed toward them, such as forcible rape.
   b. The elderly also tend to be more fearful of crime, but are the least likely to be victimized. Young men of color between the ages of 12 and 24 have the highest criminal victimization rates.
   c. The burden of robbery victimization falls more heavily on males than females, African Americans more than whites, and young people more than middle-aged and older persons.

VI. CRIMINAL JUSTICE SYSTEM
   A. The criminal justice system includes the police, the courts, and prisons. This system is a collection of bureaucracies that possesses considerable discretion -- the use of personal judgment regarding whether to take action on a situation and, if so, what kind of action to take.
   B. The **police** are responsible for crime control and maintenance of order.
   C. The **courts** determine the guilt or innocence of those accused of committing a crime.
   D. **Punishment** is any action designed to deprive a person of things of value (including liberty) because of something the person is thought to have done.
      1. Disparate treatment of the poor, people of color, and women is evident in the prison system.
      2. The **medicalization of deviance** is the transformation of deviance into a medical problem that requires treatment by a physician.
   E. For many years, capital punishment, or the death penalty, has been used in the United States; about 4,000 executions have occurred in the U.S. since 1930, and scholars have documented race and class biases in the imposition of the death penalty in this country.

VII. DEVIANCE AND CRIME IN THE TWENTY-FIRST CENTURY
   A. Although many people in the United States agree that crime is one of the most important problems facing this country, they are divided over what to do about it.
   B. The best approach for reducing delinquency and crime ultimately is prevention: to work with young people before they become juvenile offenders so as to help them establish

family relationships, build self-esteem, choose a career, and get an education which will help them pursue that career.

C.   As long as racism, sexism, classism, and ageism exist in our society, people will see deviant and criminal behavior through a selective lens.

ANALYZING AND UNDERSTANDING THE BOXES

After reading the chapter and studying the outline, re-read the four boxes and write down key points and possible questions for class discussion.

Sociology and Everyday Life -- "How Much Do You Know About Gangs, Deviance, and Crime?"

Key Points:

Discussion Questions:

1.

2.

3.

Sociology in Global Perspective -- "Street Youths, *Bosozoku*, and *Yakuza* in Japan"

Key Points:

Discussion Questions:

1.

2.

3.

Sociology and Media -- "The Media Gives Voice to 'Boys Will Be Boys'"

Key Points:

Discussion Questions:

1.

2.

3.

Sociology and Law -- "Juvenile Offenders and 'Equal Justice Under the Law'"

Key Points:

Discussion Questions:

1.

2.

3.

PRACTICE TEST

MULTIPLE CHOICE QUESTIONS

Select the response that best answers the question or completes the statement:

1.   _____ refer(s) to systematic practices developed by social groups to encourage conformity and to discourage deviance. (pp. 232-233)
    a.   Law
    b.   Folkways
    c.   Mores
    d.   Social control

2.   The text defines deviance as any: (p. 233)
    a.   aberrant behavior.
    b.   behavior, belief, or condition that violates cultural norms.
    c.   serious violation of consistent moral codes.
    d.   perverted act.

3. According to functionalists such as Emile Durkheim, deviance serves all of the following functions, except: (pp. 235-236)
   a. deviance helps us to identify social dynamite and social junk in a society.
   b. deviance clarifies rules.
   c. deviance promotes social change.
   d. deviance unites a group.

4. According to _____ theory, people are sometimes exposed to cultural goals that they are unable to obtain because they do not have access to culturally approved means of achieving those goals. (p. 236)
   a. containment
   b. status inaccessibility
   c. strain
   d. conflict

5. All of the following are included in Robert Merton's modes of adaptation to cultural goals and approved ways of achieving them, except: (pp. 236-237)
   a. retribution.
   b. ritualism.
   c. retreatism.
   d. rebellion.

6. Based on Robert Merton's typology, a government service employee who adheres to the established rules so completely that she or he often loses sight of the agency's purpose is engaged in: (p. 237)
   a. retribution.
   b. ritualism.
   c. retreatism.
   d. rebellion.

7. According to _____ theory, a teenager living in a poverty-ridden area of a central city is unlikely to become wealthy through a Harvard education, but some of his desires may be met through behaviors such as theft, drug dealing, and robbery. (p. 238)
   a. deviance management
   b. control
   c. illegitimate opportunity structures
   d. critical

8. _____ theories suggest that the probability of delinquency increases when a person's social bonds are weak and when peers promote antisocial values and violent behavior. (p. 240)
    a. Deviance management
    b. Control
    c. Illegitimate opportunity structures
    d. Critical

9. According to Edwin Lemert's typology, _____ deviance is exemplified by a person under the legal drinking age who orders an alcoholic beverage at a local bar but is not "caught" and labeled a deviant. (p. 241)
    a. primary
    b. secondary
    c. residual
    d. adolescent

10. The _____ approach argues that criminal law protects the interests of the affluent and powerful. (p. 244)
    a. functionalist
    b. liberal feminist
    c. interactionist
    d. critical

11. _____ feminism explains women's deviance and crime as a rational response to gender discrimination experienced in work, marriage, and interpersonal relationships. (p. 244)
    a. Radical
    b. Communist
    c. Liberal
    d. Socialist

12. Socialist feminists argue that women's deviance and crime occurs because: (p. 244)
    a. women are exploited by other women.
    b. women are exploited by capitalism and patriarchy.
    c. women experience gender discrimination in work, marriage, and interpersonal relationships.
    d. women are consumers and tend to purchase more than they can afford.

13. All of the following theorists are correctly matched with their theory, except: (p. 246)
    a. Howard Becker -- labeling theory
    b. Robert Merton -- strain theory
    c. Travis Hirschi -- differential association
    d. Meda Chesney-Lind -- feminist approach

14. Violent crime, certain property crimes, and certain morals crimes are referred to as ___ crime. (p. 247)
    a. misdemeanor
    b. organized
    c. political
    d. conventional

15. All of the following are index crimes, except:
    a. traffic violations.
    b. larceny in excess of $50.
    c. armed robbery.
    d. murder.

16. At the heart of much _____ crime is a violation of positions of trust in business or government. (p. 249)
    a. white-collar
    b. street
    c. organized
    d. conventional

17. Drug trafficking, prostitution, loan-sharking, and money-laundering are examples of _____ crime. (p. 250)
    a. white-collar
    b. street
    c. organized
    d. conventional

18. All of the following are examples of political crime, except: (p. 251)
    a. unethical or illegal use of government authority for the purpose of material gain.
    b. money-laundering.
    c. engaging in graft through bribery, kickbacks, or "insider" deals.
    d. dubious use of public funds and public property.

19. According to the text, rates of arrest:
    a. are about the same for males and females at every age group and for most offenses.
    b. are slightly higher for females than males in the younger age levels and for violent crimes.
    c. are about the same for males and females for prostitution due to more stringent enforcement of criminal laws pertaining to male customers.
    d. remain higher for males than females at every age and for nearly all offenses.

20. All of the following are functions of punishment, except: (p. 260)
    a. deterrence.
    b. retribution.
    c. social protection.
    d. elimination of social dynamite and social junk.

TRUE-FALSE QUESTIONS

T   F   1. People may be regarded as deviant if they express a radical or unusual belief system. (p. 234)

T   F   2. According to sociologists, deviance is relative. (p. 235)

T   F   3. Deviance is found in all societies. (p. 235)

T   F   4. Primary/secondary deviance and labeling theory are functionalist perspectives on deviance and crime. (p. 235)

T   F   5. According to differential association theory, people learn the necessary techniques and motivation for deviant behavior from people with whom they associate. (p. 240)

T   F   6. Moral entrepreneurs use their views of right and wrong to establish rules for their daily lives. (p. 241)

T   F   7. Studies of juvenile offenders show that persons from lower-income families are more likely to be arrested and indicted than are individuals from middle-income families. (p. 241)

T　　F　　8.　Gangs are virtually nonexistent in Japan, where high levels of conformity are expected. (p. 242)

T　　F　　9.　Karl Marx wrote extensively about deviance and crime. (p. 243)

T　　F　　10.　Recently, functionalist theorists have examined the relationship between class, race, and crime. (p. 244)

T　　F　　11.　Few studies of violent crime have examined the role of women as victims or perpetrators. (p. 245)

T　　F　　12.　Many people do not regard occupational and corporate crime as "criminal" behavior. (p. 250)

T　　F　　13.　Organized crime groups engage only in illegal enterprises. (p. 251)

T　　F　　14.　Official crime statistics provide very accurate data on the number of crimes committed each year in the United States. (p. 252)

T　　F　　15.　Arrest rates for index crimes are highest for people between the ages of 13 and 25. (p. 253)

SOCIOLOGY IN OUR TIMES: DIVERSITY ISSUES

1.　According to the text, "gang membership provides some women and men in low-income central-city areas with an illegitimate means to acquire money, entertainment, refuge, physical protection, and escape from living like their parents." (p. 239) Can you think of other ways society might help these young people to meet their needs other than through criminal activity?

2.　Why has research on women as victims and perpetrators of crime been extremely limited until recently? If you were going to study gender and crime, what topics might you explore?

3.　As a category, do white women and people of color have the same opportunities to commit high-level occupational and corporate crimes as white men? Does your answer tell us anything about people's access (or lack of access) to high-paying, prestigious careers and professions?

4.   Can you explain why gang membership has increased among middle-class suburban youths? Why are females more visible in gangs? Have you ever been a member of a gang? Do you think you would join a gang if you were a young suburban teenager? Would you join if you lived in a low-income central-city area?

5.   How are race, class, gender, and age interlocked with crime and punishment in the United States? Is "equal justice under the law" strictly an ideal? Is it possible for a society to have an equitable criminal justice system? Why or why not?

CHAPTER SEVEN CROSSWORD PUZZLE

For those who enjoy cross word puzzles, here is a puzzle that contains words and names from Chapter Seven. Working the puzzle will help you in reviewing the chapter. The answers appear on page 137.

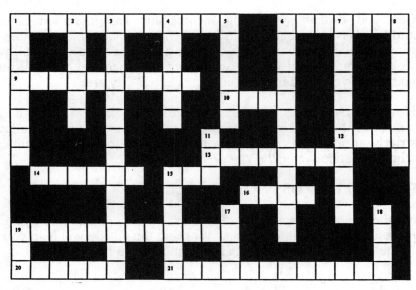

ACROSS

1. _____ association theory: individuals have a greater tendency to deviate from societal norms when they frequently associate with persons who are more favorable towards deviance

6. He and Lloyd Ohlin stated that three delinquent subcultures emerge based on the type of [19 across] opportunity structures available in a specific area

9. According to Merton, _____ occurs when people accept society's goals but adopt disapproved means for achieving them

10. According to the text, this is one example of a crime for which the punishment typically is more than a year's imprisonment

12. Official crime statistics ____ offenses that are not reported

13. The police, courts, and prisons collectively are known as the _____ justice system

14. One of the groups of high school boys that William Chambliss observed was the _____

15. Instead of stealing something like a _____, Nathan McCall's gang stole an ice cream truck

16. This is the other category compared with 8 down

19. _____ opportunity structures: circumstances that allow people to acquire through illegitimate activities what they cannot afford through legitimate channels

20. Author of the concept named in 21 across

21. _____ states that people feel strain when exposed to cultural goals that they are unable to obtain

DOWN

1. Any behavior, belief, or condition that violates cultural norms

2. A serious crime for which punishment typically ranges from more than a year's imprisonment to death

3. Function of punishment that seeks to return offenders to the community as law-abiding citizens

4. Social bond _____: the probability of deviant behavior increases when a person's ties to society are weakened or broken

5. He distinguished primary deviance from secondary deviance

6. All violent crime, certain property crimes, and certain morals crimes

7. Illegal activities committed by people in the course of their employment or financial affairs

8. From a conflict perspective, law defines and controls two categories of people, one of which is social _____

11. Initials for the major source of information on crimes committed in the U.S.

15. Property _____ include robbery, burglary, larceny, motor vehicle theft, and arson

17. One of the feminist scholars who developed theories about gender and crime

18. The other feminist scholar listed in the discussion of gender and crime

19. Suffix in two of Merton's forms of adaption

134

ANSWERS TO PRACTICE TEST, CHAPTER 7

Answers to Multiple Choice Questions

1. d Social control refers to systematic practices developed by social groups to encourage conformity and to discourage deviance. (pp. 232-233)

2. b The text defines deviance as any behavior, belief, or condition that violates cultural norms. (p. 233)

3. a According to functionalists such as Emile Durkheim, deviance serves all of the following functions, except: deviance helps us to identify social dynamite and social junk in a society. (pp. 235-236)

4. c According to strain theory, people are sometimes exposed to cultural goals that they are unable to obtain because they do not have access to culturally approved means of achieving those goals. (p. 236)

5. a All of the following are included in Robert Merton's modes of adaptation to cultural goals and approved ways of achieving them, except retribution. (pp. 236-237)

6. b Based on Robert Merton's typology, a government service employee who adheres to the established rules so completely that she or he often loses sight of the agency's purpose is engaged in ritualism. (p. 237)

7. c According to illegitimate opportunity structures theory, a teenager living in a poverty-ridden area of a central city is unlikely to become wealthy through a Harvard education, but some of his desires may be met through behaviors such as theft, drug dealing, and robbery. (p. 238)

8. b Control theories suggest that the probability of delinquency increases when a person's social bonds are weak and when peers promote antisocial values and violent behavior. (p. 240)

9. a According to Edwin Lemert's typology, primary deviance is exemplified by a person under the legal drinking age who orders an alcoholic beverage at a local bar but is not "caught" and labeled a deviant. (p. 241)

10. d The critical approach argues that criminal law protects the interests of the affluent and powerful. (p. 244)

11. c Liberal feminism explains women's deviance and crime as a rational response to gender discrimination experienced in work, marriage, and interpersonal relationships. (p. 244)

12. b Socialist feminists argue that women's deviance and crime occurs because women are exploited by capitalism and patriarchy. (p. 244)

13.   c   All of the following theorists are correctly matched with their theory, <u>except</u> Travis Hirschi -- differential association. Hirschi should be matched with social control/social bonding theory; Edwin Sutherland originated differential association theory. (p. 246).

14.   d   Violent crime, certain property crimes, and certain morals crimes are referred to as conventional crime. (p. 247)

15.   a   All of the following are index crimes, <u>except</u> traffic violations. (p. 247)

16.   a   At the heart of much white-collar crime is a violation of positions of trust in business or government. (p. 249)

17.   c   Drug trafficking, prostitution, loan-sharking, and money-laundering are examples of organized crime. (p. 250)

18.   b   All of the following are examples of political crime, <u>except</u> money-laundering. (p. 251)

19.   d   According to the text, rates of arrest remain higher for males than females at every age and for nearly all offenses. (p. 254)

20.   d   All of the following are functions of punishment, <u>except</u>: elimination of social dynamite and social junk. (p. 260)

Answers to True-False Questions

1. True (p. 234)
2. True (p. 235)
3. True (p. 235)
4. False -- Functionalist perspectives include strain theory, opportunity theory, and social control/social bonding. Primary/secondary deviance and labeling theory are interactionist perspectives. (p. 235)
5. True (p. 240)
6. False -- Moral entrepreneurs use their own views of right and wrong to establish rules and label others as deviant. (p. 241)
7. True (p. 241)
8. False -- As discussed in Box 7.2, Japanese gangs deliberately draw public attention to their deviant status. (p. 242)
9. False -- Although Marx wrote very little about deviance and crime, many of his ideas are found in a critical approach that has emerged from earlier Marxist and radical perspectives on criminology. (p. 243)
10. False -- Recently, critical conflict theorists have examined the relationship between class, race, and crime. (p. 244)
11. True (p. 245)
12. True (p. 250)
13. False -- Organized crime groups have infiltrated the world of legitimate business, such as banking, real estate, garbage collection, and garment manufacturing. (p. 251)

14. False -- Official crime statistics reflect only those crimes that have been reported to the police; victimization surveys indicate that the incidence of some crimes is substantially higher than reported in official crime reports. (p. 252)
15. True (p. 253).

ANSWER TO CHAPTER SEVEN CROSSWORD PUZZLE

```
D I F F E R E N T I A L       L         C L O W A R D
E     F     E       H         E         O       H         Y
V     L     H       E         M         N       I         N
I N N O V A T I O N     E     R A P E   V       T         A
A     N     B       R               R   E       E         M
N     Y     I       Y               T   N       C         I
C           L             U             T       O M I T
E           I             C R I M I N A L       L         E
    S A I N T S       C A R             O       L
            A       R           J U N K A       R
          T       I         M           L       R       D
I L L E G I T I M A T E     E           L               A
S           O       E       D                           L
M E R T O N         S T R A I N T H E O R Y
```

CHAPTER 8
SOCIAL STRATIFICATION AND CLASS

BRIEF CHAPTER OUTLINE
What Is Social Stratification?
Systems of Stratification
      Slavery
      The Caste System
      The Class System
Classical Perspectives on Social Class
      Karl Marx: Relation to Means of Production
      Max Weber: Wealth, Prestige, and Power
Functionalist Perspectives on the U.S. Class Structure
      Functionalist Approach to Measuring Class
      Functionalist Model of the Class Structure
      Functionalist Explanation of Social Inequality
Conflict Perspectives on the U.S. Class Structure
      Conflict Approach to Measuring Class
      Conflict Model of the Class Structure
      Conflict Explanation of Social Inequality
Inequality in the United States
      Unequal Distribution of Income and Wealth
      Consequences of Inequality
Poverty
      Who Are the Poor?
      Economic and Structural Sources of Poverty
      Solving the Poverty Problem
Social Stratification in the Twenty-First Century

CHAPTER SUMMARY
Social stratification is the hierarchical arrangement of large social groups based on their control over basic resources. A key characteristic of systems of stratification is the extent to which the structure is flexible. **Slavery**, a form of stratification in which people are owned by others, is a closed system. In a **caste system** people's status is determined at birth based on their parents' position in society. The **class system**, which exists in the United States, is a type of stratification based on ownership of resources and on the type of work people do. Karl Marx and Max Weber acknowledged social class as a key determinant of social inequality and social change. According to Marx, capitalistic societies are comprised of two classes -- the capitalists, who own the means of production, and the workers, who sell their labor to the owners. By contrast, Weber developed a multidimensional concept that focuses on the

interplay of **wealth, prestige,** and **power.** Functionalist perspectives on the U.S. class structure view classes as broad groupings of people who share similar levels of privilege based on their roles in the occupational structure. According to the Davis-Moore thesis, positions that are most important within society, requiring the most talent and training, must be highly rewarded. Conflict perspectives are based on the assumption that social stratification is created and maintained by one group in order to enhance and protect its own economic interests. The stratification of society into different social groups results in wide discrepancies in income, wealth, and access to available goods and services (including health, good nutrition, and education). Sociologists distinguish between **absolute poverty,** which exists when people do not have the means to secure the basic necessities of life, and **relative poverty,** which exists when people may be able to afford basic necessities but are still unable to maintain an average standard of living. There are both economic and structural sources of poverty. Low wages are a key problem, as are unemployment and underemployment. As the gap between rich and poor, employed and unemployed widens, social inequality will increase in the twenty-first century if we do nothing. Given that the well-being of all people is linked, it is incumbent that we ensure that everyone has a job, a living wage, and an equal life chance.

LEARNING OBJECTIVES
After reading Chapter 8, you should be able to:
1. Define social stratification and describe the major sources of stratification found in societies.

2. Explain social mobility and distinguish between intergenerational and intragenerational mobility.

3. Describe the key characteristics of the three major systems of stratification.

4. Describe Karl Marx's perspective on class position and class relationships.

5. Outline Max Weber's multidimensional approach to social stratification and explain how people are ranked on all three dimensions.

6. Compare functionalist and conflict approaches to measuring class.

7. Outline the functionalist model of the U.S. class structure and briefly describe the key characteristics of each class.

8. Distinguish between functionalist and conflict explanations of social inequality.

9. Outline the conflict model of the U.S. class structure and briefly describe the key characteristics of each class.

10. Discuss the distribution of income and wealth in the United States and describe how this distribution affects life chances.

11. Distinguish between absolute and relative poverty and describe the characteristics and lifestyle of those who live in poverty in the United States.

12. Describe the feminization of poverty and explain why two out of three impoverished adults in the United States are women.

KEY TERMS (defined at page number shown and in glossary)

absolute poverty 298
apartheid 276
basic class location 288
caste system 276
class system 276
contradictory class locations 288
feminization of poverty 299
intergenerational mobility 275
intragenerational mobility 275
invidious distinctions 272
job deskilling 301

life chances 273
meritocracy 287
pink collar occupations 285
power 280
prestige 280
relative poverty 298
slavery 275
social mobility 275
social stratification 272
socioeconomic status (SES) 282
wealth 280

KEY PEOPLE (identified at page number shown)

John Butler 289
Patricia Hill Collins 283
Kingsley Davis and
  · Wilbert Moore 286
Dennis Gilbert and
  Joseph A. Kahl 283

Karl Marx 278
Diana Pearce 299
Max Weber 280
Eric Olin Wright 288

CHAPTER OUTLINE

I. WHAT IS SOCIAL STRATIFICATION?
   A. **Social stratification** is the hierarchical arrangement of large social groups based on their control over basic resources.
   B. Max Weber's term **life chances** describes the extent to which persons within a particular layer of stratification have access to important scarce resources.

II. SYSTEMS OF STRATIFICATION
   A. Systems of stratification may be open or closed based on the availability of **social mobility** -- the movement of individuals or groups from one level in a stratification system to another.
      1. **Intergenerational mobility** is the social movement experienced by family members from one generation to the next.
      2. **Intragenerational mobility** is the social movement of individuals within their own lifetime
   B. **Slavery**, a closed system, is an extreme form of stratification in which some people are owned by others.
   C. A **caste system** is a system of social inequality in which people's status is permanently determined at birth based on their parents' ascribed characteristics.
   D. The **class system** is a type of stratification based on the ownership and control of resources and on the type of work people do.

III.   CLASSICAL PERSPECTIVES ON SOCIAL CLASS
      A.   Karl Marx: Relation to Means of Production
            1.   According to Marx, class position in capitalistic societies is determined by people's work situation, or relationship to the means of production.
                  a.   The bourgeoisie or capitalist class consists of those who privately own the means of production; the proletariat, or workers, must sell their labor power to the owners in order to earn enough money to survive.
                  b.   Class relationships involve inequality and exploitation; workers are exploited as capitalists expropriate a surplus value from their labor.
            2.   The capitalist class maintained its position by control of the society's superstructure -- comprised of the government, schools, and other social institutions which produce and disseminate ideas perpetuating the existing system.
      B.   Max Weber: Wealth, Prestige, and Power
            1.   Weber's multidimensional approach to stratification focused on the interplay among wealth, prestige, and power as being necessary in determining a person's class position.
                  a.   Weber placed people who have a similar level of **wealth** -- the value of all of a person's or family's economic assets, including income, personal property, and income-producing property -- and income in the same class.
                  b.   **Prestige** is the respect or regard with which a person or status position is regarded by others, and those who share similar levels of social prestige belong to the same status group regardless of their level of wealth.
                  c.   **Power** -- the ability of people or groups to carry out their own goals despite opposition from others -- gives some people the ability to shape society in accordance with their own interests and to direct the actions of others.
            2.   Wealth, prestige, and power are separate continuums on which people can be ranked from high to low; individuals may be high on one dimension while being low on another.

IV. FUNCTIONALIST PERSPECTIVES ON THE U.S. CLASS STRUCTURE
   A. Functionalist Approach to Measuring Class
       1. Three methods of measuring class include: the subjective approach -- people are asked to locate themselves in the class structure; the reputational approach -- people are asked to place other individuals in their community (based on their reputation) into social classes; and the objective approach -- researchers assign individuals to social classes based on predetermined criteria (e.g., occupation, income, education).
       2. **Socioeconomic status (SES)** -- a combined measure that attempts to classify individuals, families, or households in terms of indicators such as income, occupation, and education -- is used to determine class location.
   B. Functionalist Model of the Class Structure
       1. The Upper (or Capitalist) Class is the wealthiest and most powerful class, comprised of people who own substantial income-producing assets.
       2. The Upper-Middle Class is based on a combination of three factors: university degrees, authority and independence on the job, and high income. Examples of occupations for this class are highly educated professionals such as physicians, stockbrokers, or corporate managers.
       3. The Middle Class is characterized by a minimum of a high school diploma or a community college degree.
       4. The Working Class is comprised of semiskilled machine operatives, clerks and salespeople in routine, mechanized jobs, and workers in **pink collar occupations** -- relatively low-paying, nonmanual semiskilled positions primarily held by women.
       5. The Working Poor live from just above to just below the poverty line; they hold unskilled jobs, seasonal migrant employment in agriculture, lower-paid factory jobs, and service jobs (e.g., such as counter help at restaurants).
       6. The Underclass includes people who are poor, seldom employed, and caught in long term deprivation.

C.   Functionalist Explanation of Social Inequality
   1. According to the Davis-Moore thesis:
      a. All societies have important tasks that must be accomplished and certain positions that must be filled.
      b. Some positions are more important for the survival of society than others.
      c. The most important positions must be filled by the most qualified people.
      d. The positions that are the most important for society and require scarce talent, extensive training, or both, must be the most highly rewarded.
      e. The most highly rewarded positions should be those which are functionally unique (no other position can perform the same function), and those positions upon which others rely for expertise, direction, or financing.
   2. This thesis assumes that social stratification results in **meritocracy** -- a hierarchy in which all positions are rewarded based on people's ability and credentials.

V.   CONFLICT PERSPECTIVES ON THE U.S. CLASS STRUCTURE
   A. Conflict Approach to Measuring Class
      1. Erik Olin Wright outlined four criteria for placement in the class structure: (a) ownership of the means of production; (b) purchase of the labor of others (employing others); (c) control of the labor of others (supervising others on the job); and (d) sale of one's own labor (being employed by someone else).
         a. According to Wright, **basic class location** -- positions in the class structure where issues of property ownership and control are relatively clear -- is determined by factors of ownership and authority.
         b. **Contradictory class location** refers to positions within the productive process that possess a combination of elements from two different basic class locations.
   B. A Conflict Model of the Class Structure
      1. The Capitalist Class is composed of those who have inherited fortunes, own major corporations, or are top corporate executives who own extensive amounts of stock or control company investments.
      2. The Managerial Class includes upper-level managers -- supervisors and professionals who typically do not

participate in companywide decisions -- and lower-level managers who may be given some control over employment practices, such as the hiring and firing of some workers.

    3.   The Small-Business Class consists of small business owners, craftspeople, and some doctors and lawyers who may hire a small number of employees but largely do their own work.

    4.   The Working Class is made up of blue-collar workers, including skilled workers (e.g., electricians, plumbers, and carpenters), unskilled blue-collar workers (e.g., laundry and restaurant workers), and white-collar workers do not own the means of production, do not control the work of others, and are relatively powerless in the workplace.

  C.   A Conflict Explanation of Social Inequality

    1.   From a conflict perspective, inequality does not serve as a source of motivation for people; powerful individuals and groups use ideology to maintain their favored positions at the expense of others.

    2.   Core values, laws, and informal social norms support inequality in the United States (e.g., legalized segregation and discrimination produce higher levels of economic inequality).

VI.   INEQUALITY IN THE UNITED STATES

  A.   Income and wealth are very unevenly distributed in the United States.

    1.   **Income** is the economic gain derived from wages, salaries, income transfers (governmental aid such as AFDC), or ownership of property.

    2.   **Wealth** includes not only income but also property such as buildings, land, farms, houses, factories, cars, and other assets.

  B.   Consequences of Inequality

    1.   Health and Nutrition: As people's economic status increases so does their health status; the poor have shorter life expectancies and are at greater risk for chronic illnesses and infectious diseases. About 40 million people in the United States are without health insurance coverage.

    2.   Education and life chance are directly linked; while functionalists view education as an "elevator" for social mobility, conflict theorists stress that schools are agencies for reproducing the capitalist class system and perpetuating inequality in society.

VII. POVERTY
   A.   Although some people living in poverty are unemployed, many hardworking people with full-time jobs also live in poverty.
   B.   The official poverty line is based on what is considered to be the minimum amount of money required for living at a subsistence level.
   C.   Sociologists distinguish between **absolute poverty** -- when people do not have the means to secure the most basic necessities of life -- and **relative poverty** -- when people may be able to afford basic necessities but still are unable to maintain an average standard of living.
   D.   Who Are the Poor?
        1.   Age: Children are more likely to be poor than older persons; older women are twice as likely to be poor as older men; older African Americans and Latinos/as are much more likely to live below the poverty line than are non-Latino/a whites.
        2.   Gender: About two-thirds of all adults living in poverty are women; this problem is described as the **feminization of poverty** -- the trend in which women are disproportionately represented among individuals living in poverty.
        3.   Race and Ethnicity: white Americans (non-Latinos/as) account for approximately two-thirds of those below the official poverty line; however, a disproportionate percentage of the poverty population is made up of African Americans, Latinos/as, and Native Americans.
   E.   Economic and Structural Sources of Poverty
        1.   An economic source of poverty is the low wages paid for many jobs: Half of all families living in poverty are headed by someone who is employed, and one-third of those family heads work full time.
        2.   Poverty also is exacerbated by structural problems such as (a) deindustrialization -- millions of U.S. workers have lost jobs as corporations have disinvested here and opened facilities in other countries where "cheap labor" exists -- and (b) **job deskilling** -- a reduction in the proficiency needed to perform a specific job that leads to a corresponding reduction in the wages paid for that job.
   F.   Solving the Poverty Problem
        1.   The United States has attempted to solve the poverty problem with social welfare programs; however, the primary beneficiaries have not always been the poor.

146

2. A lack of consensus exists regarding both the definition of the problem and the possible solutions for it.

VIII. THE TWENTY-FIRST CENTURY AND SOCIAL STRATIFICATION

    A. According to some social scientists, wealth will become more concentrated at the top of the U.S. class structure; as the rich have grown richer, more people have found themselves among the ranks of the poor.

    B. Structural sources of upward mobility are shrinking while the rate of downward mobility has increased; the persistence of economic inequality is related to profound global economic changes.

ANALYZING AND UNDERSTANDING THE BOXES

After reading the chapter and studying the outline, re-read the four boxes and write down key points and possible questions for class discussion.

Sociology and Everyday Life -- "How Much Do You Know About Wealth, Poverty, and the American Dream?"

Key Points:

Discussion Questions:

1.

2.

3.

Sociology in Global Perspective -- "Slavery in Brazil Today"

Key Points:

Discussion Questions:

1.

2.

3.

147

Sociology and Media -- "Wealth and the American Dream"

Key Points:

Discussion Questions:

1.

2.

3.

Sociology and Law -- "Social Welfare and the American Dream"

Key Points:

Discussion Questions:

1.

2.

3.

PRACTICE TEST

MULTIPLE CHOICE QUESTIONS

Select the response that best answers the question or completes the statement:

1. _____ refers to wide discrepancies in the income, wealth, life conditions, life chances, and lifestyles between people as a result of systems of stratification. (p. 272)
    a. Absolute poverty
    b. Relative poverty
    c. Invidious distinctions
    d. Vidious distinctions

2. Sociologists use the term _____ to refer to the hierarchial arrangements of large social groups based on their control over basic resources. (p. 272)
    a.   social stratification
    b.   social layering
    c.   social distinction
    d.   social accumulation

3. A young woman's father is a carpenter; she graduates from college with a degree in accounting, becomes a CPA, and has a starting salary that represents more money than her father ever made in one year. This illustrates _____ mobility. (p. 275)
    a.   intragenerational
    b.   intergenerational
    c.   horizontal
    d.   subjective

4. All of the following are true statements about slavery, except: (p. 275)
    a.   Slavery is a closed system in which "slaves" are treated as property.
    b.   Slaves were forcibly imported to the United States as a source of cheap labor.
    c.   Slavery has ended throughout the world.
    d.   Some people have been enslaved because of unpaid debts, criminal behavior, or war and conquest.

5. A _____ system is a system of social inequality in which people's status is permanently determined at birth based on their parents' ascribed characteristics. (p. 276)
    a.   class
    b.   slavery
    c.   capitalist
    d.   caste

6. A young man who comes from an impoverished background works at two full-time jobs in order to save enough money to attend college. Ultimately, he earns a degree, attends law school, and is hired by a firm at a starting salary of $60,000. This person has experienced _____ mobility. (p. 276)
    a.   horizontal
    b.   vertical
    c.   direct
    d.   indirect

7. According to _____, class position is determined by people's relationship to the means of production. (p. 278)
    a. Karl Marx
    b. Max Weber
    c. Emile Durkheim
    d. Dennis Gilbert and Joseph A. Kahl

8. All of the following statements regarding Marx's analysis of class are correct, except: (pp. 278-279)
    a. Class relationships involve inequality and exploitation.
    b. The exploitation of workers by the capitalist class ultimately will lead to the destruction of capitalism.
    c. The capitalist class maintains its position by control of the society's superstructure.
    d. Wealth, prestige, and power are separate continuums on which people can be ranked from high to low.

9. According to Max Weber, _____ is the respect or regard with which a person or status position is regarded by others. (p. 280)
    a. admiration
    b. power
    c. prestige
    d. rank

10. In Max Weber's stratification typology, white-collar workers, public officials, managers, and professionals make up the _____ class. (p. 280)
    a. upper
    b. middle
    c. working
    d. lower

11. A researcher instructs respondents to place other individuals in their community into social classes, based upon the respondents' perceptions of how these individuals are thought of in the community. The investigator is utilizing the _____ method. (p. 281)
    a. ethnomethodological
    b. objective
    c. subjective
    d. reputational

12. Prestige ratings for selected occupations in the United States have shown that the occupation of _____ consistently receives the highest ranking. (p. 282)
   a. attorney
   b. physician
   c. accountant
   d. dentist

13. All of the following are limitations of status attainment research models, except: (pp. 282-283)
   a. this research is based on sophisticated statistical measures.
   b. this research focuses on traditionally male jobs and excludes women's work.
   c. this research has limited use in explaining African American class dynamics.
   d. this research assumes that upward mobility is available to everyone in the United States.

14. According to the functionalist model of the class structure, members of the _____ class have earned most of their money in their own lifetime as entrepreneurs, presidents of corporations, top-level professionals, and so forth. (p. 283)
   a. upper-upper
   b. lower-upper
   c. upper-middle
   d. middle

15. The text points out that a combination of three factors qualifies people for the upper-middle class. Which of the following is not one of these factors? (p. 284)
   a. university degrees
   b. authority and independence on the job
   c. inherited wealth
   d. high income

16. Over the past fifty years, Asian Americans, Latinos/as, and African Americans have placed great emphasis on _____ as a means of attaining the American Dream. (p. 284)
   a. education
   b. affirmative action
   c. unemployment compensation
   d. vocational training

17. According to the functionalist explanation of social inequality: (p. 286)
   a. all societies have important tasks that must be accomplished and certain positions that must be filled.
   b. the most important positions must be filled by the most qualified people.
   c. the most highly rewarded positions should be those that are functionally unique and on which other positions rely.
   d. all of the above.

18. According to the conflict explanation of social inequality: (p. 291)
   a. the existence of social inequality serves as a motivating force for people.
   b. the wealthy are smarter than other people.
   c. laws and informal social norms support inequality in the United States.
   d. all of the above.

19. All of the following are included in Erik O. Wright's typology of the class structure, except the _____ class. (p. 288)
   a. capitalist
   b. managerial
   c. middle
   d. working

20. The trend in which women disproportionately are represented among individuals living in poverty is referred to as _____: (p. 299)
   a. absolute poverty.
   b. relative poverty.
   c. situational poverty.
   d. the feminization of poverty.

TRUE-FALSE QUESTIONS

T   F   1. According to Max Weber, lifestyle describes the extent to which persons within a particular layer of stratification have access to important scarce resources. (p. 273)

T   F   2. Race/ethnicity, gender, and religion affect people's social mobility. (p. 278)

T   F   3. Sociologists refer to relatively low-paying, nonmanual, semiskilled positions primarily held by women as blue-collar occupations. (p. 285)

T   F   4.   The Davis-Moore thesis assumes that social stratification results in meritocracy. (p. 287)

T   F   5.   According to Erik O. Wright, basic class location refers to positions in the class structure in which issues of property ownership and control are relatively clear. (p. 288)

T   F   6.   Core values in the United States support the prevailing resource distribution system and contribute to social inequality. (p. 291)

T   F   7.   Among prosperous nations, Japan is number one in inequality of income distribution. (p. 293)

T   F   8.   For most people in the United States, wealth is invested primarily in property that generates no income, such as a home or car. (p. 294)

T   F   9.   As people's economic status increases, so does their health status. (p. 296)

T   F   10.   About 40 million people in the United States are without health insurance coverage. (p. 296)

T   F   11.   According to conflict theorists, schools are agencies for perpetuating poverty. (p. 297)

T   F   12.   Relative poverty exists when people do not have the means to secure the most basic necessities of life. (p. 298)

T   F   13.   About two-third of all adults living in poverty are men. (p. 299)

T   F   14.   Most of the poor and virtually all welfare recipients are people of color. (p. 300)

T   F   15.   Half of all families living in poverty are headed by someone who is employed, and one-third of those family heads work full time. (p. 301)

SOCIOLOGY IN OUR TIMES: DIVERSITY ISSUES

1. If you had $250,000 a year to spend (like George at the beginning of Chapter 8), how would you allocate the money? If you had $800 a month ($9,600 a year) to spend (like Joey at the beginning of Chapter 8), how would you allocate the money? Would the presence of children make a significant difference in the way you allocate the $250,000? How about the $9,600?

2. Do you believe in the American Dream? If so, what factors do you think will help you reach your dream? What factors do you think may limit your opportunities to reach that dream?

3. Why has education been a major factor in the upward mobility of Asian Americans, Latinos/as, and African Americans over the past fifty years? (p. 284) To what extent do you think education will help you to achieve your own goals?

4. How do recent movies and television shows portray wealth and poverty? Are race/ethnicity and gender intertwined with these class depictions?

5. According to the text, "poverty is everyone's problem." (p. 300) Do you agree with this statement? Why or why not?

CHAPTER EIGHT CROSSWORD PUZZLE

For those who enjoy crossword puzzles, here is a puzzle that contains words and names from Chapter Eight. Working the puzzle will help you in reviewing the chapter. The answers appear on page 158.

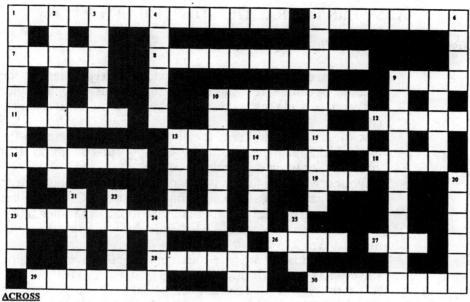

ACROSS

1. Social _____ is the hierarchal arrangement of large social groups based on their control over 2 down
5. _____ poverty exists when people do not have the means to secure the most basic necessities of life
7. _____ system: system of social inequality in which status is permanently determined at birth
8. _____: a type of stratification based on ownership and control and on the type of work people do
9. The upper-upper class does not have to ____ for a living
10. The respect or regard with which a person or status position is regarded by others
11. The "new rich" have _____ most of their money in their own lifetime
12. The American _____: we can move beyond our origins and become more successful than our parents
13. Thesis bearing his and 22-down's names states that inequality is not only inevitable but also necessary for the smooth functioning of society
15. First name of person who coined the term "life chances"
16. The people named in 30 across constitute _____ of 1% of households but own 75% of the nation's wealth
17. The extent to which persons within a particular layer of stratification have access to important scarce resources are called ____ chances
18. Most people in the U.S. think of themselves as neither ____ nor poor, but rather middle class
19. Members of the lower-upper class are the ___ rich
23. Hierarchy in which all positions are rewarded based on people's ability and credentials
24. According to some stereotypes, most of the poor and virtually all welfare recipients can be identified on the basis of _____
26. Power is the ability to [29 across] a ____ despite opposition

27. ___ deskilling: reduction in a position's required proficiency with a corresponding reduction in wages
28. In a _____ system of stratification, boundaries between levels are rigid
29. See 26 across
30. See 16 across

DOWN

1. _____ status: a combined measure that attempts to classify individuals, families, or households in terms of indicators that include income
2. Basic _____
3. Poverty in the U.S. is highly concentrated according to ____ _____ and race/ethnicity of people
4. The value of all of a person's or family's economic assets, including _____, personal property, and income-producing property is the answer to 20 down
5. Status _____ research focuses on the process by which people ultimately reach their position in the class structure
6. ____ Olin Wright outlined four criteria for placement in the class structure
9. This 20% of the U.S. population live from just above to just below the poverty line
10. From a conflict perspective, education perpetuates _____
13. _____ Pearce coined the term "the feminization of poverty"
14. Person who is owned by others
20. See 4 down
21. Vestiges of a closed system of stratification in which status is determined at _____ exist in India and South Africa
22. See 13 across
25. The research in 5 down uses the father's occupation and the ____ education and first job as determinants

155

ANSWERS TO PRACTICE TEST, CHAPTER 8

Answers to Multiple Choice Questions

1. c   Invidious distinctions refers to wide discrepancies in the income, wealth, life conditions, life chances, and lifestyles between people as a result of systems of stratification. (p. 272)

2. a   Sociologists use the term social stratification to refer to the hierarchial arrangements of large social groups based on their control over basic resources. (p. 272)

3. b   A young woman's father is a carpenter; she graduates from college with a degree in accounting, becomes a CPA, and has a starting salary that represents more money than her father ever made in one year. This illustrates intergenerational mobility. (p. 275)

4. c   All of the following are <u>true</u> statements about slavery, <u>except</u>: slavery has ended throughout the world. (p. 275)

5. d   A caste system is a system of social inequality in which people's status is permanently determined at birth based on their parents' ascribed characteristics. (p. 276)

6. b   A young man who comes from an impoverished background works at two full-time jobs in order to save enough money to attend college. Ultimately, he earns a degree, attends law school, and is hired by a firm at a starting salary of $60,000. This person has experienced vertical mobility. (p. 276)

7. a   According to Karl Marx, class position is determined by people's relationship to the means of production. (p. 278)

8. d   All of the following statements regarding Marx's analysis of class are correct, <u>except</u>: Wealth, prestige, and power are separate continuums on which people can be ranked from high to low. (p. 278-279)

9. c   According to Max Weber, prestige is the respect or regard with which a person or status position is regarded by others. (p. 280)

10. b   In Max Weber's stratification typology, white-collar workers, public officials, managers, and professionals make up the middle class. (p. 280)

11. d   A researcher instructs respondents to place other individuals in their community into social classes, based upon the respondents' perceptions of how these individuals are thought of in the community. The investigator is utilizing the reputational method. (p. 281)

12.   b   Prestige ratings for selected occupations in the United States have shown that the occupation of physician consistently receives the highest ranking. (p. 282)

13.   a   All of the following are limitations of status attainment research models, except: this research is based on sophisticated statistical measures. (pp. 282-283)

14.   b   According to the functionalist model of the class structure, members of the lower-upper class have earned most of their money in their own lifetime as entrepreneurs, presidents of corporations, top-level professionals, and so forth. (p. 283)

15.   c   The text points out that a combination of three factors qualifies people for the upper-middle class. Which of the following is not one of these factors? inherited wealth (p. 284)

16.   a   Over the past fifty years, Asian Americans, Latinos/as, and African Americans have placed great emphasis on education as a means of attaining the American Dream. (p. 284)

17.   d   According to the functionalist explanation of social inequality all of these statements are true: all societies have important tasks that must be accomplished and certain positions that must be filled; the most important positions must be filled by the most qualified people; and the most highly rewarded positions should be those that are functionally unique and on which other positions rely. (p. 286)

18.   c   According to the conflict explanation of social inequality, laws and informal social norms support inequality in the United States. (p. 291)

19.   c   All of the following are included in Erik O. Wright's typology of the class structure, except the middle class. (p. 288)

20.   d   The trend in which women are disproportionately represented among individuals living in poverty is referred to as the feminization of poverty. (p. 299)

Answers to True-False Questions

1.   False -- According to Max Weber, life chances describes the extent to which persons within a particular layer of stratification have access to important scarce resources. (p. 273)

2.   True (p. 278)

3.   False -- Sociologists refer to relatively low-paying, nonmanual, semiskilled positions primarily held by women as pink-collar occupations. (p. 285)

4.   True (p. 287)

5.   True (p. 288)

6.   True (p. 291)

7. False -- Among prosperous nations, the United States is number one in inequality of income distribution. (p. 293)
8. True (p. 294)
9. True (p. 296)
10. True (p. 296)
11. True (p. 297)
12. False -- Absolute poverty exists when people do not have the means to secure the most basic necessities of life. (p. 298)
13. False -- About two-thirds of all adults living in poverty are women. (p. 299)
14. False -- According to some stereotypes, most of the poor and virtually all welfare recipients are people of color. However, this stereotype is false; white Americans (non-Latinos/as) account for approximately two-thirds of those below the official poverty line. (p. 300)
15. True (p. 301)

ANSWER TO CHAPTER EIGHT CROSSWORD PUZZLE

```
S T R A T I F I C A T I O N     A B S O L U T E
O   E   H       I               T             R
C A S T E     C L A S S S Y S T E M         I
I   O   A       O               A       W O R K
O   U   G       M     P R E S T I G E   O   E
E A R N E D     E     O         N   D R E A M
C   C         D A V I S         M A X   K   C
O N E H A L F   I   E   L I F E   R I C H
N   S           A   R   A       N E W   N   W
O     B   M     N   T   V       T       G   E
M E R I T O C R A C Y   E   S       P   A
I     R   O   A       H   G O A L   J O B   L
C     T   R   C L O S E D   N       O   T
    A C H I E V E       R       S U P E R R I C H
```

BRIEF CHAPTER OUTLINE
Race and Ethnicity
     Social Significance of Race and Ethnicity
     Racial Classification and the Meaning of Race
     Majority and Minority Groups
Prejudice
     Theories of Prejudice
     Measuring Prejudice
Discrimination
Sociological Perspectives on Race and Ethnic Relations
     Interactionist Perspectives
     Functionalist Perspectives
     Conflict Perspectives
Racial and Ethnic Groups in the United States
     Native Americans
     White Anglo-Saxon Protestants/British Americans
     African Americans
     White Ethnics
     Asian Americans
     Latinos/as (Hispanic Americans)
     Middle Easterners
Global Racial and Ethnic Inequality in the Twenty-first Century
     Worldwide Racial and Ethnic Struggles
     Growing Racial and Ethnic Diversity in the United States

CHAPTER SUMMARY
Issues of race and ethnicity permeate all levels of interaction in the United
States. A **race** is a category of people who have been singled out as inferior
or superior, often on the basis of physical characteristics such as skin color,
hair texture, and eye shape. By contrast, an **ethnic group** is a collection of
people distinguished, by others or by themselves, primarily on the basis of
cultural or nationality characteristics. Race and ethnicity are ingrained in our
consciousness and often form the basis of hierarchical ranking and determine
who gets what resources. A **majority** (or **dominant**) **group** is one that is
advantaged and has superior resources and rights in a society while a
minority (or **subordinate**) **group** is one whose members, because of physical
or cultural characteristics, are disadvantaged and subjected to unequal
treatment by the dominant group and who regard themselves as objects of
collective discrimination. **Prejudice** is a negative attitude based on faulty

generalizations about the members of selected racial and ethnic groups. **Discrimination** -- actions or practices of dominant group members that have a harmful impact on members of a subordinate group -- may be either **individual** or **institutional discrimination** -- involving day-to-day practices of organizations and institutions that have a harmful impact on members of subordinate groups. According to the interactionist contact hypothesis, increased contact between people from divergent groups should lead to favorable attitudes and behavior when a specific set of criteria are met. Two functionalist perspectives -- **assimilation** and **ethnic pluralism** -- focus on how members of subordinate groups become a part of the mainstream. Alternately, conflict theories analyze economic stratification and access to power in race and ethnic relations: caste and class perspectives, **internal colonialism**, **split-labor market** theory, gendered racism, and racial formation theory. The unique experiences of Native Americans, White Anglo-Saxon Protestants/British Americans, African Americans, White Ethnics, Asian Americans, Latinos/as (Hispanic Americans), and Middle Easterners are discussed, and the increasing racial-ethnic diversity of the United States is examined.

LEARNING OBJECTIVES
After reading Chapter 9, you should be able to:

1. Define race and ethnic group and explain their social significance.

2. Explain the sociological usage of majority group and minority group and note why these terms may be misleading.

3. Discuss prejudice and explain the major theories of prejudice.

4. Discuss discrimination and distinguish between individual and institutional discrimination.

5. Describe interactionist perspectives on racial and ethnic relations.

6. Distinguish between assimilation and ethnic pluralism and explain why both are functionalist perspectives on racial and ethnic relations.

7. Explain the key assumptions of conflict perspectives on racial and ethnic relations and note the group(s) to which each applies.

8. Trace the intergroup relationships of racial and ethnic groups in the United States.

9. Explain how the experiences of Native Americans have been different from those of other racial and ethnic groups in the United States.

10. Describe how the African American experience in the United States has been unique when compared with other groups.

11. Compare and contrast the experiences of Chinese Americans, Japanese Americans, Korean Americans, Filipino Americans, and Indochinese Americans in the United States.

12. Describe the experiences of Mexican Americans, Puerto Ricans, and Cuban Americans in the United States.

13. Discuss racial and ethnic struggles from a global perspective.

KEY TERMS (defined at page number shown and in glossary)

assimilation 321
authoritarian personality 318
discrimination 318
ethnic group 310
ethnic pluralism 323
genocide 319
individual discrimination 319
institutional discrimination 319
internal colonialism 324
lynching 333

majority (dominant) group 315
minority (subordinate)
   group 315
prejudice 315
race 310
racism 315
scapegoat 318
segregation 323
social distance 318
split labor market 327

KEY PEOPLE (identified at page number shown)

Robert Blauner 324
Emory Bogardus 318
Joe R. Feagin 319
Robert Merton 318

Michael Omi and
   Howard Winant 328
William Julius Wilson 324

CHAPTER OUTLINE

I.   RACE AND ETHNICITY

    A.   A **race** is a category of people who have been singled out as inferior or superior, often on the basis of physical characteristics such as skin color, hair texture, and eye shape.

    B.   An **ethnic group** is a collection of people distinguished, by others or by themselves, primarily on the basis of cultural or nationality characteristics.

    C.   Social significance of race and ethnicity: Race and ethnicity are bases of hierarchical ranking in society; the dominant group holds power over other (subordinate) ethnic groups.

    D.   Racial classifications in the U.S. census mirror how the meaning of race has continued to change over the past century in the U.S.

    E.   A **majority** (or **dominant**) **group** is one that is advantaged and has superior resources and rights in a society; a **minority** (or **subordinate**) **group** is one whose members, because of physical or cultural characteristics, are disadvantaged and subjected to unequal treatment by the dominant group and who regard themselves as objects of collective discrimination.

II.   PREJUDICE

    A.   **Prejudice** is a negative attitude based on faulty generalizations about members of selected racial and ethnic groups. Prejudice is often based on **stereotypes** -- overgeneralizations about the appearance, behavior, or other characteristics of all members of a category. The frustration - aggression hypothesis states that people who are frustrated in their efforts to achieve a highly desired goal will respond with a pattern of aggression toward a **scapegoat** -- a person or group that is incapable of offering resistance to the hostility or aggression of others.

    B.   **Racism** is the belief that some racial or ethnic groups are superior while others are inferior.

    C.   Theories of prejudice include the frustration-aggression hypothesis, social learning theory, and the theory of the **authoritarian personality,** which is characterized by excessive conformity, submissiveness to authority, intolerance, insecurity, a high level of superstition, and rigid, stereotypic thinking.

    D.   Based on the work of Emory Bogardus, **social distance** is the extent to which people are willing to interact and establish relationships with members of racial and ethnic groups other than their own.

III.   DISCRIMINATION

    A.   **Discrimination** is defined as actions or practices of dominant group members that have a harmful impact on members of a subordinate group.

    B.   Robert Merton identified four combinations of attitudes and responses:

        1.   Unprejudiced nondiscriminators -- persons who are not personally prejudiced and do not discriminate against others;

        2.   Unprejudiced discriminators -- persons who may have no personal prejudice but still engage in discriminatory behavior because of peer-group prejudice or economic, political, or social interests;

        3.   Prejudiced nondiscriminators -- persons who hold personal prejudices but do not discriminate due to peer pressure, legal demands, or a desire for profits;

        4.   Prejudiced discriminators -- persons who hold personal prejudices and actively discriminate against others.

    C.   Discriminatory actions vary in severity from the use of derogatory labels to violence against individuals and groups.

        1.   **Genocide** is the deliberate, systematic killing of an entire people or nation.

        2.   More recently, the term "ethnic cleansing" has been used to define a policy of "cleansing" geographic areas (such as in Bosnia-Herzegovina) by forcing persons of other races or religions to flee -- or die.

D.   Discrimination also varies in how it is carried out.
   1. **Individual discrimination** consists of one-on-one acts by members of the dominant group that harm members of the subordinate group or their property.
   2. **Institutional discrimination** is the day-to-day practices of organizations and institutions that have a harmful impact on members of subordinate groups.

IV. SOCIOLOGICAL PERSPECTIVES ON RACE AND ETHNIC RELATIONS
   A.   Interactionist Perspectives
      1. The contact hypothesis suggests that contact between people from divergent groups should lead to favorable attitudes and behavior when a specific set of criteria is met.
      2. However, scholars have found that increasing contact may have little or no effect on existing prejudices.
   B.   Functionalist Perspectives
      1. **Assimilation** is a process by which members of subordinate racial and ethnic groups become absorbed into the dominant culture.
      2. **Ethnic pluralism** is the coexistence of a variety of distinct racial and ethnic groups within one society.
   C.   Conflict Perspectives
      1. The **caste perspective** views racial and ethnic inequality as a permanent feature of U.S. society.
      2. **Class perspectives** emphasize the role of the capitalist class in racial exploitation.
      3. **Internal colonialism** occurs when members of a racial or ethnic group are conquered, or colonized, and forcibly placed under the economic and political control of the dominant group.
      4. **Split labor market** refers to the division of the economy into two areas of employment, a primary sector composed of higher-paid (usually dominant group) workers in more secure jobs, and a secondary sector comprised of lower-paid (often subordinate group) workers in jobs with little security and frequently hazardous working conditions.
      5. **Gendered racism** refers to the interactive effect of racism and sexism in the exploitation of women of color.
      6. The **theory of racial formation** states that actions of the government substantially define racial and ethnic relations in the United States.

V. RACIAL AND ETHNIC GROUPS IN THE UNITED STATES
A. Native Americans
1. Historically, Native Americans experienced the following kinds of treatment in the United States:
a. Genocide
b. Forced Migration
c. Forced Assimilation
2. Today, about two million Native Americans live in the United States (primarily in the southwest), and about one-third live on reservations.
3. Native Americans are the most disadvantaged racial or ethnic group in the United States in terms of income, employment, housing, nutrition, and health (especially among individuals living on reservations).
B. White Anglo-Saxon Protestants/British Americans
1. Although many English settlers initially were indentured servants or sent here as prisoners, they quickly emerged as the dominant group, creating a core culture to which all other groups were expected to adapt.
2. Like other racial and ethnic groups, British Americans are not all alike; social class and gender affect their life chances and opportunities.
C. African Americans
1. Slavery was rationalized by stereotyping African Americans as inferior and childlike; however, some slaves and whites engaged in active resistance that eventually led to the abolition of slavery.
2. Through informal practices in the north and Jim Crow laws in the south, African Americans experienced segregation in housing, employment, education, and all public accommodations.
3. **Lynching** -- a killing carried out by a group of vigilantes seeking revenge for an actual or imagined crime by the victim -- was used by whites to intimidate African Americans into staying "in their place."
4. During World Wars I and II, African Americans were a vital source of labor in war production industries; however, racial discrimination continued both on and off the job.
a. After African Americans began to demand sweeping societal changes in the 1950s, racial segregation slowly was outlawed by the courts and the federal government.

165

        b.   Civil rights legislation attempted to do away with discrimination in education, housing, employment, and health care.

    5.   Today, African Americans make up about 13 percent of the U.S. population; many have made significant gains in education, employment, and income in the past three decades; however, other African Americans have not fare so well; for example, the African American unemployment rate remains twice as high as that of whites, and young people in central city areas face a bleak future.

D.   The term **white ethnics** was coined to identify immigrants who came from European countries other than England: Ireland, Poland, Italy, Greece, Germany, Yugoslavia, Russia and other former Soviet republics, and so forth.

E.   Asian Americans

    1.   Chinese Americans

          a.   The initial wave of Chinese immigration occurred between 1850 and 1880 when Chinese men came to the United States seeking gold in California and jobs constructing the transcontinental railroads.

          b.   Chinese Americans were subjected to extreme prejudice and stereotyping; the Chinese Exclusion Act of 1882 was passed because white workers feared for their jobs.

          c.   In the 1960s, the second and largest wave of Chinese immigration came from Hong Kong and Taiwan.

          d.   Today, one-third of all Chinese Americans were born in the United States; as a group, they have enjoyed considerable upward mobility, but many Chinese Americans live in poverty in Chinatowns.

    2.   Japanese Americans

          a.   The earliest Japanese immigrants primarily were men who worked on sugar plantations in the Hawaiian Islands in the 1860s; the immigration of Japanese men was curbed in 1908; however, Japanese women were permitted to enter the U.S. for several more years because of the shortage of women.

          b.   Internment: During World War II, when the United States was at war with Japan, nearly 120,000 Japanese Americans were placed in internment camps because they were seen as a security threat; many Japanese Americans lost all that they owned during the internment.

    c.   In spite of the extreme hardship faced as a result of the loss of their businesses and homes during World War II, many Japanese Americans have been very successful.

3.   Korean Americans

    a.   The first wave of Korean immigrants were male workers who arrived in Hawaii between 1903 and 1910; the second wave came to the mainland following the Korean War in 1964 (e.g., the wives of servicemen, and Korean children who had lost their parents in the war); and the third wave arrived after the Immigration Act of 1965 permitted well-educated professionals to migrate to the U.S.

    b.   Korean Americans have helped each other open small businesses by pooling money through the kye -- an association that grants members money on a rotating basis to gain access to more capital.

4.   Filipino Americans

    a.   Most of the first Filipino immigrants were men who were employed in agriculture; following the Immigration Act of 1965, Filipino physicians, nurses, technical workers, and other professionals moved in large numbers to the U.S. mainland.

    b.   Unlike other Asian Americans, most Filipinos have not had the start-up capital necessary to open their own businesses, and workers generally have been employed in the low-wage sector of the dual labor market.

5.   Indochinese Americans

    a.   Most Indochinese Americans (including people from Vietnam, Cambodia, Thailand, and Laos) have come to the U.S. in the past two decades.

    b.   Vietnamese refugees who had the resources to flee at the beginning of the Vietnam War were the first to arrive. Next came Cambodians and lowland Laotians, referred to as "boat people" by the media.

        c.   Today, most Indochinese Americans are foreign born; about half live in western states, especially California. Even though most Indochinese immigrants spoke no English when they arrived in this country, some of their children have done very well in school and have been stereotyped as "brains".

F.   Latinos/as (Hispanic Americans)

    1.   Mexican Americans or Chicanos/as have experienced disproportionate poverty as a result of internal colonialism.

        a.   More recently, Mexican Americans have been seen as cheap labor at the same time that they have been stereotyped as lazy.

        b.   When anti-immigration sentiments are running high, Mexican Americans often are the objects of discrimination.

        c.   Today, the families of many Mexican Americans have lived in the United States for four or five generations and have made significant contributions in many areas.

    2.   When Puerto Rico became a possession of the United States in 1917, Puerto Ricans acquired U.S. citizenship and the right to move freely to and from the mainland; while living conditions have improved substantially for some, others have continued to live in poverty in Spanish Harlem and other barrios.

    3.   Cuban Americans have fared somewhat better than other Latinos; early waves of Cuban immigrants were affluent business and professional people; the second wave of Cuban Americans in the Mariel boatlift of the 1970s fared worse; and more recent arrivals have developed their own ethnic and economic enclaves in cities such as Miami.

G.   Middle Easterners

    1.   Since 1970, many immigrants have arrived in the United States from Middle Eastern countries such as Egypt, Syria, Lebanon, Iran, and Jordan.

    2.   While some are from working class families, the Lebanese, Syrians, and Iranians primarily come from middle class backgrounds.

    3.   Most Iranian immigrants initially hoped to return to Iran; however, many now have become U.S. citizens and are creating their own ethnic enclaves.

VI. GLOBAL RACIAL AND ETHNIC INEQUALITY IN THE TWENTY-FIRST CENTURY
    A. Worldwide Racial and Ethnic Struggles
        1. The cost of self-determination -- the right to choose one's own way of life -- often is the loss of life and property in ethnic warfare (e.g., Bosnia and Herzegovina, Croatia, Spain, Romania, Russia, Moldova, Georgia, the Middle East, Africa, Asia, and Latin America).
        2. However, some analysts predict that the "superpower" nations, including the United States, Great Britain, Japan, and Germany, will suppress ethnic violence with the assistance of the United Nations, which will serve a peacekeeping function by monitoring and enforcing agreements between rival factions.
    B. Growing Racial and Ethnic Diversity in the United States
        1. Racial and ethnic diversity is increasing in the United States: by the year 2000, white Americans will make up 70 percent of the population, in contrast to 80 percent in 1980; by 2056, the roots of the average U.S. resident will be Africa, Asia, Hispanic countries, the Pacific Islands, or Arabia -- not white Europe.
        2. Interethnic tensions may ensue between whites and people of color; people may continue to employ sincere fictions -- personal beliefs that are a reflection of larger societal mythologies, such as "I am not a racist" -- even when these are inaccurate perceptions.
        3. Some analysts believe that there is reason for cautious optimism; throughout U.S. history subordinate racial and ethnic groups have struggled to gain the freedom and rights which were previously withheld from them, and movements comprised of both whites and people of color will continue to oppose racism in everyday life, to aim at healing divisions among racial groups, and to teach children about racial tolerance.

ANALYZING AND UNDERSTANDING THE BOXES

    After completing the fill-in study outline, re-read the four boxes and write down key points and possible questions for class discussion.

Sociology and Everyday Life -- "How Much Do You Know About Race, Ethnicity, and Sports?"

Key Points:

Discussion Questions:

1.

2.

3.

Sociology in Global Perspective -- "Racial and Ethnic Stacking on Sports Teams"

Key Points:

Discussion Questions:

1.

2.

3.

Sociology and Media -- "Be Like Shaq? The Marketing of African American Athletes"

Key Points:

Discussion Questions:

1.

2.

3.

Sociology and Law -- "Everyone Should Own a Professional Team"

Key Points:

Discussion Questions:

1.

2.

3.

PRACTICE TEST

MULTIPLE CHOICE QUESTIONS

Select the response that best answers the question or completes the statement:

1.   All of the following are characteristics of ethnic groups, except: (p. 310)
   a.   unique cultural traits.
   b.   a feeling of ethnocentrism.
   c.   territoriality.
   d.   the same religion.

2.   According to the text, the terms "majority group" and "minority group" are: (p. 315)
   a.   accurate because "majority groups" always are larger in number than "minority groups."
   b.   misleading because people who share ascribed racial or ethnic characteristics automatically constitute a group.
   c.   less accurate terms than "dominant group" and "subordinate group," which more accurately reflect the importance of power in the relationships.
   d.   no longer used because they are "politically incorrect."

3.   _____ is a negative attitude based on faulty generalizations about members of selected racial and ethnic groups. (p. 315)
   a.   Prejudice
   b.   Discrimination
   c.   Stereotyping
   d.   Genocide

4. According to the text, the use of Native American names, images, and mascots by sports teams is an example of: (p. 315)
   a. prejudice.
   b. discrimination.
   c. stereotyping.
   d. genocide.

5. "Stacking" refers to: (p. 316)
   a. the assignment of students of color to schools where their racial-ethnic group is in the majority.
   b. the assignment of players to positions on sports teams based on ascribed characteristics rather than achieved characteristics.
   c. pent-up frustrations that cause people to unleash their feelings of frustration on a scapegoat.
   d. the spatial and social separation of categories of people by race, ethnicity, class, gender, and/or religion.

6. All of the following are examples of overt racism, except:
   a. a police detective who repeatedly refers to African Americans in derogatory terms while discussing police department procedures with the author of a screenplay.
   b. stacking.
   c. segregation of students by race/ethnicity in public schools.
   d. hate crimes perpetrated against victims because of their race/ethnicity, sexual orientation, or religion.

7. Sociologist Emory Bogardus developed a scale to measure: (p. 318)
   a. racial-ethnic prejudice in society.
   b. stacking.
   c. social distance.
   d. income differentials by race/ethnicity.

8. According to sociologist Robert Merton's typology, an umpire who is prejudiced against African Americans and deliberately makes official calls against them when they are at bat, is an example of a(n): (p. 319)
   a. unprejudiced nondiscriminator.
   b. unprejudiced discriminator.
   c. prejudiced nondiscriminator.
   d. prejudiced discriminator.

9. Institutional discrimination consists of: (p. 319)
   a. one-on-one acts by members of the dominant group that harm members of the subordinate group or their property.
   b. day-to-day practices of organizations and institutions that have a harmful impact on members of subordinate groups.
   c. the division of the economy into two areas of employment, a primary sector or upper tier, and a secondary sector or lower tier.
   d. the deliberate, systematic killing of an entire people or nation.

10. Special education classes that originally were intended to provide extra educational opportunities for children with various types of disabilities but now amount to a form of racial segregation in many school districts are an example of _____ discrimination. (p. 320)
    a. indirect institutionalized
    b. direct institutionalized
    c. small-group
    d. isolate

11. According to the contact hypothesis, contact between people from divergent groups should lead to favorable attitudes and behaviors when certain factors are present. Members of each group must have all of the following, except: (pp. 320-321)
    a. equal status.
    b. pursuit of the same goals.
    c. being competitive with one another to achieve their goals.
    d. positive feedback from interactions with each other.

12. _____ is a process by which members of subordinate racial and ethnic groups become absorbed into the dominant culture. (p. 321)
    a. Assimilation
    b. Ethnic pluralism
    c. Accommodation
    d. Internal colonialism

13. _____ is the coexistence of a variety of distinct racial and ethnic groups within one society. (p. 323)
    a. Assimilation
    b. Ethnic pluralism
    c. Accommodation
    d. Internal colonialism

14. All of the following theories are conflict perspectives on racial-ethnic relations, except: (p. 323)
    a.   the caste perspective.
    b.   split-labor market theory.
    c.   internal colonialism.
    d.   ethnic pluralism.

15. Based on early theories of race relations by W. E. B. Du Bois, sociologist Oliver C. Cox suggested that African Americans were enslaved because: (p. 324)
    a.   of prejudice based on skin color.
    b.   they were the cheapest and best workers that owners could find for heavy labor in the mines and on plantations.
    c.   they allowed themselves to be dominated by wealthy people.
    d.   they agreed to work for passage to the United States, believing that they would be released after a period of indentured servitude.

16. Sociologist _____ has suggested that class is more important than race in explaining "black-white relations" in the United States. (p. 324)
    a.   William Julius Wilson
    b.   Joe R. Feagin
    c.   Oliver C. Cox
    d.   Robert Merton

17. According to _____, sports reflects the interests of the wealthy and powerful; athletes are exploited in order for coaches, managers, and owners to gain high levels of profit and prestige. (p. 324)
    a.   interactionists
    b.   sportsologists
    c.   functionalists
    d.   conflict theorists

18. The division of the economy into a primary sector, composed of higher-paid workers in more secure jobs, and a secondary sector, composed of lower-paid workers in jobs with little security and hazardous working conditions is referred to as: (p. 327)
    a.   the split-labor market.
    b.   racial formation.
    c.   gendered racism.
    d.   stacking.

19. The experiences of Native Americans in the United States have been characterized by: (p. 329)
    a. slavery and segregation.
    b. restrictive immigration laws.
    c. genocide, forced migration, and forced assimilation.
    d. internment.

20. Of all Latino/a categories residing in the United States, _have experienced disproportionate poverty as a result of internal colonialism. (p. 339)
    a. Cuban Americans.
    b. Puerto Ricans.
    c. recent immigrants from the Dominican Republic.
    d. Mexican Americans.

TRUE-FALSE QUESTIONS

T   F   1. Although members of ethnic groups over time may become indistinguishable from the dominant group, racial groups have greater difficult blending into the mainstream, if they choose to do so. (p. 310)

T   F   2. Racial classifications in the U.S. census have remained unchanged for the past century. (p. 313)

T   F   3. The terms "majority group" and "minority group" are misleading because, numerically speaking, "minority" means a group that is fewer in number than a "majority" group. (p. 315)

T   F   4. Analysts have found that whites who accept racial stereotypes have no greater desire for social distance from people of color than do whites who reject negative stereotypes. (p. 318)

T   F   5. The extermination of 6 million European Jews by Nazi Germany is an example of genocide. (p. 319)

T   F   6. Discrimination is an individual activity; it is never based on norms of an organization or community. (p. 320)

T   F   7. Equalitarian pluralism is an ideal typology that has never actually been identified in any society. (p. 323)

T   F   8.   The class perspective views racial and ethnic inequality as a permanent feature of U. S. society. (p. 343)

T   F   9.   The internal colonialism model helps explain the continued exploitation of immigrant groups such as the Chinese, Filipinos, and Vietnamese. (p. 327)

T   F   10.   The term "gendered racism" refers to the interactive effect of racism and sexism in the exploitation of women of color. (p. 327)

T   F   11.   White Anglo-Saxon Protestants (WASPS) have been the most privileged group in the United States. (p. 331)

T   F   12.   The term "white ethnics" was coined to identify immigrants who came from European countries other than England, such as Ireland, Poland, Germany, and Italy. (p. 335)

T   F   13.   White ethnics have not experienced discrimination in the United States. (p. 336)

T   F   14.   Most of the first Chinese, Japanese, Korean, and Filipino immigrants in the United States were men. (pp. 337-338).

T   F   15.   It is predicted that by 2056, the roots of the average U.S. resident will be in Africa, Asia, Hispanic countries, the Pacific Islands, and the Middle East. (p. 341)

SOCIOLOGY IN OUR TIMES: DIVERSITY ISSUES

1.   Do you think Arthur Ashe's statement at the beginning of Chapter 9 accurately reflects the impact of race and ethnicity on people's lives in the United States, or do you think it is an overreaction on his part? Does your answer mirror your own racial-ethnic experiences in the United States and/or other countries?

2.   The next time you watch a sports event on television or in person, think about the race/ethnicity of the team members. Does your observation support (or refute) the ideas in this chapter? What part

does gender play in sports events? Where are the men, and what are they doing? Where are the women, and what are they doing?

3. If we are truly honest, most of us hold at least some slight feelings of prejudice against persons we consider to be different from ourselves. But many of us do not overtly discriminate against people we have defined as being in our "outgroups." Do you think it is possible to build higher levels of trust and tolerance in highly diverse societies such as the United States?

4. To what extent does the "political climate" in a country affect racial-ethnic relations? Can members of the U.S. Congress and the president bring about greater racial harmony? Can they also exacerbate racial divisiveness, strife, and social inequalities?

CHAPTER NINE CROSSWORD PUZZLE

For those who enjoy crossword puzzles, here is a puzzle that contains words and names from Chapter Nine. Working the puzzle will help you in reviewing the chapter. The answers appear on page 181.

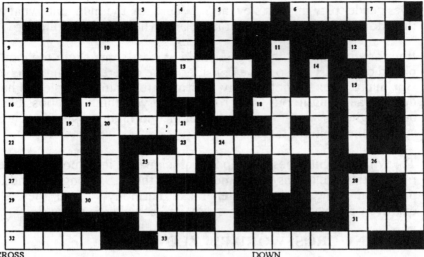

ACROSS

1. Actions or practices of [1 down] group members (or their representatives) that have a harmful [5 down] on members of a subordinate group

6. The belief that some racial or ethnic groups are superior while others are [23 across]

9. According to _____ and Howard Winant, race "permeates every institution, every relationship, and every individual"

12. Category of people who have been singled out as inferior or superior, often on the basis of physical characteristics

13. Ku Klux ____: organization known for its prejudice against minority groups

15. ____ that working-class whites were losing jobs to Chinese immigrants resulted [17 across] the Chinese Exclusion Act of 1882

16. See 4 down

17. See 15 across

18. See 7 down

20. Elena Albert described the experience of hearing Marcus Garvey and realizing she did not have to be like the _____ of white womanhood

22. The [32 across] labor market refers to the division of the economy into two areas of employment, one of which is the primary sector or upper ____

23. See 6 across

25. An example of stereotypes in sports is the team name "Washington ___ skins"

26. If racial equality existed in the U.S., race would ___ be a consideration in the selection of leaders, hiring decisions, etc.

29. When a person is seen as conforming to a stereotype, he or she may be treated simply as one of "___ people"

30. A killing carried by a group of vigilantes seeking revenge

31. See 27 down

32. See 22 across

33. A collection of people distinguished primarily on the basis of cultural or nationality characteristics

DOWN

1. See 1 across

2. _____ distance: the extent to which people are willing to interact and establish relationships with members of racial and ethnic groups other than their own

3. _____ discrimination: harmful action intentionally taken by a dominant group member against a member of a subordinate group

4. Like members of other racial and ethnic groups, not [16 across] WASPs are _____

5. See 1 across

7. First part of word [with 18 across] meaning a person or group that is incapable of offering resistance to the hostility or aggression of others

8. The spatial and social separation of categories of people by race, ethnicity, class, gender, and/or religion

10. While the terms race and _____ sometimes are used interchangeably, some sociologists believe the impact of the former is greater than the impact of the latter

11. The deliberate, systematic killing of an entire people or nation

14. _____ colonialism: occurs when members of a racial or ethnic group are conquered or colonized and forcibly placed under the control of the dominant group

19. Jim ____ laws legalized separation of the races in all public accommodations

21. U.S. immigration policies have been designed to keep the "___" on certain types of immigrants

24. Sociologist who identified four major types of discrimination

27. Among the physical characteristics on which some people single out a race as superior or inferior are [31 across] color, hair texture, and ____ shape

28. Abbreviated name for British Americans

178

ANSWERS TO PRACTICE TEST, CHAPTER 9

Answers to Multiple Choice Questions

1. d   All of the following are characteristics of ethnic groups, except: the same religion. (p. 310)

2. c   According to the text, the terms "majority group" and "minority group" are less accurate terms than "dominant group" and "subordinate group," which more accurately reflect the importance of power in the relationships. (p. 315)

3. a   Prejudice is a negative attitude based on faulty generalizations about members of selected racial and ethnic groups. (p. 315)

4. c   According to the text, the use of Native American names, images, and mascots by sports teams is an example of stereotyping. (p. 315)

5. b   "Stacking" refers to the assignment of players to positions on sports teams based on ascribed characteristics rather than achieved characteristics. (p. 316)

6. b   All of the following are examples of overt racism, except stacking (p. 316).

7. c   Sociologist Emory Bogardus developed a scale to measure social distance. (p. 318)

8. d   According to sociologist Robert Merton's typology, an umpire who is prejudiced against African Americans and deliberately makes official calls against them when they are at bat, is an example of a prejudiced discriminator. (p. 319)

9. b   Institutional discrimination consists of day-to-day practices of organizations and institutions that have a harmful impact on members of subordinate groups. (p. 319)

10. a   Special education classes that originally were intended to provide extra educational opportunities for children with various types of disabilities but now amount to a form of racial segregation in many school districts are an example of indirect institutionalized discrimination. (p. 320)

11. c   According to the contact hypothesis, contact between people from divergent groups should lead to favorable attitudes and behaviors when certain factors are present. Members of each group must have all of the following, except being competitive with one another to achieve their goals. (pp. 320-321)

12. a   Assimilation is a process by which members of subordinate racial and ethnic groups become absorbed into the dominant culture. (p. 321)

13. b   Ethnic pluralism is the coexistence of a variety of distinct racial and ethnic groups within one society. (p. 323)

14. d   All of the following theories are conflict perspectives on racial-ethnic relations, except: ethnic pluralism. (p. 323)

15. b Based on early theories of race relations by W. E. B. Du Bois, sociologist Oliver C. Cox suggested that African Americans were enslaved because they were the cheapest and best workers that owners could find for heavy labor in the mines and on plantations. (p. 324)

16. a Sociologist William Julius Wilson has suggested that class is more important than race in explaining "black-white relations" in the United States. (p. 324)

17. d According to conflict theorists, sports reflects the interests of the wealthy and powerful; athletes are exploited in order for coaches, managers, and owners to gain high levels of profit and prestige. (p. 324)

18. a The division of the economy into a primary sector, composed of higher-paid workers in more secure jobs, and a secondary sector, composed of lower-paid workers in jobs with little security and hazardous working conditions is referred to as the split-labor market. (p. 327)

19. c The experiences of Native Americans in the United States have been characterized by genocide, forced migration, and forced assimilation. (p. 329)

20. d Of all Latino/a categories residing in the United States, Mexican Americans have experienced disproportionate poverty as a result of internal colonialism. (p. 339)

Answers to True-False Questions
1. True (p. 310)
2. False -- Racial classifications in the U.S. census have been modified over the past century; however, this classification system remains outdated and inadequate, as explained in the text (p. 313).
3. True (p. 315)
4. False -- Analysts have found that whites who accept racial stereotypes <u>do</u> have a greater desire for social distance from people of color than do whites who reject negative stereotypes. (p. 318)
5. True (p. 319)
6. False -- Although discriminatory actions are engaged in by individuals, some organizations and communities do prescribe actions that intentionally have a differential and negative impact on members of subordinate groups. (p. 320)
7. False -- Some analysts suggest that Switzerland is an example of equalitarian pluralism because over 6 million people with French, German, and Italian cultural heritages peacefully coexist there. (p. 323)
8. False -- The **caste** perspective views racial and ethnic inequality as a permanent feature of U. S. society. (p. 343)

9. False -- The internal colonialism model <u>does</u> <u>not</u> help explain the continued exploitation of immigrant groups such as the Chinese, Filipinos, and Vietnamese. It helps to explain the experiences of indigenous populations (such as Native Americans and Mexican Americans) who were colonized by Euro-Americans and others who invaded their lands and conquered them. (p. 327)
10. True (p. 327)
11. True (p. 327)
12. True (p. 335)
13. False -- White ethnics <u>have</u> experienced discrimination in the United States. See text for examples. (p. 336)
14. True (pp. 337-338)
15. True (p. 341)

ANSWER TO CHAPTER NINE CROSSWORD PUZZLE

D	I	S	C	R	I	M	I	N	A	T	I	O	N		R	A	C	I	S	M	
O		O				S		L		M								C		S	
M	I	C	H	A	E	L	O	M	I		P		G			R	A	C	E		
I		I			T		L		K	L	A	N		E		I		D		E	
N		A			H		A		E		C		N		N		F	E	A	R	
A	L	L		I	N		T			T		G	O	A	T		A			E	
N		C		I	D	E	A	L				C		E		N				G	
T	I	E	R		C			I	N	F	E	R	I	O	R		S			A	
		O		I	:	R	E	D		E		D		N				N	O	T	
E		W		T		I			A			E		A		W				I	
Y	O	U		L	Y	N	C	H	I	N	G			L		A				O	
E						O				I						S	K	I	N		
S	P	L	I	T				E	T	H	N	I	C	G	R	O	U	P		

CHAPTER 10
SEX AND GENDER

BRIEF CHAPTER OUTLINE

Sex and Gender
    Sex
    Gender
    The Social Significance of Gender
    Sexism
Gender Stratification in Historical Perspective
    Hunting and Gathering Societies
    Horticultural and Pastoral Societies
    Agrarian Societies
    Industrial Societies
Gender and Socialization
    Gender Socialization by Parents
    Peers and Gender Socialization
    Teachers and Schools and Gender Socialization
    Sports and Gender Socialization
    Mass Media and Gender Socialization
    Adult Gender Socialization
Contemporary Gender Inequality
    Gendered Division of Paid Work
    Pay Equity, or Comparable Worth
    Paid Work and Family Work
Perspectives on Gender Stratification
    Functionalist and Neoclassical Economic Perspectives
    Conflict Perspectives
    Feminist Perspectives
Gender Issues in the Twenty-First Century

CHAPTER SUMMARY

It is important to distinguish between **sex** -- the biological and anatomical differences between females and males -- and **gender** -- the culturally and socially constructed differences between females and males found in the meanings, beliefs, and practices associated with "femininity" and "masculinity." Gender is socially significant because it leads to differential treatment of men and women; sexism (like racism) often is used to justify discriminatory treatment. Sexism is linked to patriarchy, a hierarchical system in which cultural, political, and economic structures are male dominated. In most hunting and gathering societies, fairly equitable relationships exist because neither sex has the ability to provide all of the food necessary for survival. In horticultural societies, hoe cultivation is compatible with child care, and a fair degree of gender equality exists because neither sex controls the food supply. In agrarian societies, male dominance is very apparent; tasks require more labor and physical strength, and women are seen as too weak or too tied to child-rearing activities to perform these activities. In industrialized societies, a gap exists between unpaid work performed by women at home and paid work performed by men and women. The key agents of gender socialization are parents, peers, teachers and schools, sports, and the mass media, all of which tend to reinforce stereotypes of appropriate gender behavior. Gender inequality results from the economic, political, and educational discrimination of women. In most workplaces, jobs are either gender segregated or the majority of employees are of the same gender. Gender segregated occupations lead to a disparity, or pay gap, between women's and men's earnings. Even when women are employed in the same job as men, on average they do not receive the same, or comparable, pay. Many women have a "second shift" because of their dual responsibilities for paid and unpaid work. According to functional analysts, husbands perform instrumental tasks of economic support and decision making, and wives assume expressive tasks of providing affection and emotional support for the family. Conflict analysts suggest that the gendered division of labor within families and the workplace results from male control and dominance over women and resources. Although feminist perspectives vary in their analyses of women's subordination, they all advocate social change to eradicate gender inequality.

LEARNING OBJECTIVES
After reading Chapter 10, you should be able to:
1. Distinguish between sex and gender and explain their sociological significance.

2. Explain why sex is not always clear-cut; differentiate between hermaphrodites, transsexuals, and transvestites.

3. Describe the relationship between gender roles, gender identity, and body consciousness.

4. Define sexism and explain how it is related to discrimination and patriarchy.

5. Trace gender stratification from early hunting and gathering societies until today.

6. Describe the process of gender socialization and identify specific ways in which parents, peers, teachers, sports, and mass media contribute to the process.

7. Describe gender bias and explain how schools operate as a gendered institution.

8. Discuss the gendered division of paid work and explain its relationship to the issue of pay equity or comparable worth.

9. Trace changes in labor force participation by women and note how these changes have contributed to the "second shift."

10. Describe functionalist and neoclassical economic perspectives on gender stratification and contrast them with conflict perspectives.

11. State the feminist perspective on gender equality and outline the key assumptions of liberal, radical, socialist, and black (African American) feminism.

KEY TERMS (defined at page number shown and in glossary)

body consciousness 352

comparable worth 373

feminism 378

gender 352

gender bias 365

gender identity 352

gender role 352

hermaphrodite 351

matriarchy 354

patriarchy 354

pay gap 372

primary sex characteristics 350

secondary sex characteristics 351

sex 349

sexism 354

sexual orientation 351

transsexual 351

transvestite 351

KEY PEOPLE (identified at page number shown)

Ben Agger 378

Susan Bordo 354

Paula England and
   Melissa Herbert 374

George F. Gilder 376

Elizabeth Higginbotham 370

Dorothy C. Holland and
   Margaret A. Eisenhart 364

Alice Kemp 377

Jean Kilbourne 369

Judith Lorber 352, 370

Myra Sadker and
   David Sadker 365

Becky W. Thompson 353

CHAPTER OUTLINE

I.   SEX AND GENDER

   A.   **Sex** refers to the biological and anatomical differences between females and males.

      1.   **Primary sex characteristics** are the genitalia used in the reproductive process; **secondary sex characteristics** are the physical traits (other than reproductive organs) that identify an individual's sex.

      2.   **Sexual orientation** is a preference for emotional-sexual relationships with members of the opposite sex (heterosexuality), the same sex (homosexuality), or both (bisexuality).

   B.   **Gender** refers to the culturally and socially constructed differences between females and males found in the meanings, beliefs, and practices associated with "femininity" and "masculinity."

      1.   A microlevel analysis of gender focuses on how individuals learn **gender roles** -- the attitudes, behavior, and activities that are socially defined as appropriate for each sex and are learned through the socialization process -- and **gender identity** -- a person's perception of the self as female or male.

      2.   A macrolevel analysis of gender examines structural features, external to the individual, which perpetuate

gender inequality, including gendered institutions, that are reinforced by a gendered belief system, based on ideas regarding masculine and feminine attributes that are held to be valid in a society.

C. **Sexism** -- the subordination of one sex, usually female, based on the assumed superiority of the other sex -- is interwoven with **patriarchy** -- a hierarchical system of social organization in which cultural, political, and economic structures are controlled by men.

II. GENDER STRATIFICATION IN HISTORICAL PERSPECTIVE

A. The earliest known division of labor between women and men is in hunting and gathering societies.

B. In horticultural societies, women make an important contribution to food production because hoe cultivation is compatible with child care; a fairly high degree of gender equality exists because neither sex controls the food supply.

C. In pastoral societies, herding primarily is done by men; women contribute relatively little to subsistence production and thus have relatively low status.

D. Gender inequality increases in agrarian societies as men become more involved in food production.

E. In industrial societies -- those in which factory or mechanized production has replaced agriculture as the major form of economic activity -- the status of women tends to decline further.

F. Gendered division of labor increases the economic and political subordination of women.

III. GENDER AND SOCIALIZATION

A. Parents as Agents of Gender Socialization

   1. From birth, parents act toward children on the basis of gender labels; children's clothing and toys reflect their parents' gender expectations.

   2. Boys are encouraged to engage in gender-appropriate behavior; they are not to show an interest in "girls'" activities.

B. Peers and Gender Socialization

   1. Peers help children learn prevailing gender-role stereotypes, as well as gender-appropriate and inappropriate behavior.

   2. During adolescence, peers often are stronger and more effective agents of gender socialization than are adults.

   3. Among college students, peers play an important role in career choices and the establishment of long term, intimate relationships.

186

C.   Teachers and Schools and Gender Socialization
1.   From kindergarten through college, schools operate as gendered institutions; teachers provide important messages about gender through both the formal content of classroom assignments and informal interactions with students.
2.   Teachers may unintentionally demonstrate **gender bias** -- the showing of favoritism toward one gender over the other -- toward male students.

E.   Sports and Gender Socialization
1.   The type of game played differs with the child's sex: from elementary school through high school, boys play football and other competitive sports while girls are cheerleaders, members of the drill team, and homecoming queens.
2.   For many males, sports participation and spectatorship is a training ground for masculinity; for females, sports still is tied to the male gender role, thus making it very difficult for girls and women to receive the full benefits of participating in such activities.

F.   Mass Media and Gender Socialization
1.   Gender stereotyping is found in media, ranging from children's cartoons to adult shows.
2.   On television, more male than female roles are shown, and male characters typically are more aggressive, constructive, and direct, while females are deferential toward others or use manipulation to get their way.
3.   Advertising also plays an important role in gender socialization.

G.   Adult Gender Socialization
1.   Men and women are taught gender-appropriate conduct in schools and the workplace.
2.   Different gender socialization may occur as people reach their forties and enter "middle age."

IV.   CONTEMPORARY GENDER INEQUALITY
A.   Gender-segregated work refers to the concentration of women and men in different occupations, jobs, and places of work.
B.   Labor market segmentation -- the division of jobs into categories with distinct working conditions -- results in women having separate and unequal jobs in the secondary sector of the split- or dual-labor market that are lower paying, less prestigious, and have fewer opportunities for advancement.
C.   Occupational segregation contributes to a **pay gap** -- the disparity between women's and men's earnings. **Comparable**

worth is the belief that wages ought to reflect the worth of a job, not the gender or race of the worker.

D.   Although both men and women profess that working couples should share household responsibilities, researchers find that family demands remain mostly women's responsibility, even among women who hold full-time paid employment.

V.   PERSPECTIVES ON GENDER STRATIFICATION

A.   Functional and neoclassical economic perspectives on the family view the division of family labor as ensuring that important societal tasks will be fulfilled.

1.   According to the human capital model, individuals vary widely in the amount of education and job training they bring to the labor market.

2.   From this perspective, what individuals earn is the result of their own choices and labor market demand for certain kinds of workers at specific points in time.

3.   Other neoclassical economic models attribute the wage gap to such factors as: (1) the different amounts of energy men and women expend on their work; (2) the occupational choices women make (choosing female-dominated occupations so that they can spend more time with their families); and (3) the crowding of too many women into some occupations (suppressing wages because the supply of workers exceeds demand).

B.   According to the conflict perspective, the gendered division of labor within families and the workplace results from male control of and dominance over women and resources.

1.   Although men's ability to use physical power to control women diminishes in industrial societies, they still remain the head of household, control the property, and hold more power through their predominance in the most highly paid and prestigious occupations and the highest elected offices.

2.   Conflict theorists in the Marxist tradition assert that gender stratification results from private ownership of the means of production; some men not only gain control over property and the distribution of goods but also gain power over women.

C.   Feminist Perspectives

1.   **Feminism** refers to a belief that women and men are equal and that they should be valued equally and have equal rights.

2.   In **liberal feminism**, gender equality is equated with equality of opportunity.

3. According to **radical feminists**, male domination causes all forms of human oppression, including racism and classism.

4. **Socialist feminists** suggest that women's oppression results from their dual roles as paid and unpaid workers in a capitalist economy. In the workplace, women are exploited by capitalism; at home, they are exploited by patriarchy.

5. **Black (African American) feminism** is based on the belief that women of color experience a different world than other people because of multilayered oppression based on race/ethnicity, gender, and class.

VI. GENDER ISSUES IN THE TWENTY-FIRST CENTURY

A. In the past 30 years, women have made significant progress in the labor force; laws have been passed to prohibit sexual discrimination in the workplace and school; women are more visible in education, the workplace, and government.

B. Many men have joined feminist movements not only to raise their consciousness about men's concerns, but also about the need to eliminate sexism and gender bias.

C. However, U.S. society still remains far from gender equity in many areas of life.

ANALYZING AND UNDERSTANDING THE BOXES

After reading the chapter and studying the outline, re-read the four boxes and write down key points and possible questions for class discussion.

Sociology and Everyday Life -- "How Much Do You Know About Body Image and Gender?"

Key Points:

Discussion Questions:

1.

2.

3.

Sociology in Global Perspective -- "Women and Human Rights: Female Genital Mutilation"

Key Points:

Discussion Questions:

1.

2.

3.

Sociology and Media -- "You've Come a Long Way, Baby!"

Key Points:

Discussion Questions:

1.

2.

3.

Sociology and Law -- Weight-Based Discrimination

Key Points:

Discussion Questions:

1.

2.

3.

PRACTICE TEST

MULTIPLE CHOICE QUESTIONS

Select the response that best answers the question or completes the statement:

1. In responding to the question, "Why do women and men feel differently about their bodies?" the text points out that: (p. 348)
   a. cultural differences in appearance norms may explain women's greater concern.
   b. in the United States, the image of female beauty as childlike and thin is continually flaunted by the advertising industry.
   c. women of all racial-ethnic groups, classes, and sexual orientations regard their weight as a crucial index of their acceptability to others.
   d. all of the above.

2. Sociologists use the term _____ to refer to the biological attributes of men and women; _____ is used to refer to the distinctive qualities of men and men that are culturally created. (p. 349)
   a. gender - sex
   b. sex - gender
   c. primary sex characteristics -- secondary sex characteristics
   d. secondary sex characteristics -- primary sex characteristics

3. A(n) _____ is a person in whom sexual differentiation is ambiguous or incomplete. (p. 351)
   a. hermaphrodite
   b. transsexual
   c. transvestite
   d. homosexual

4. A person's perception of the self as female or male is referred to as one's: (p. 352)
   a. self concept.
   b. gender identity.
   c. gender role.
   d. gender belief system.

5. According to historian Susan Bordo, the anorexic body and the muscled body: (p. 354)
   a. illustrate that eating problems and bodybuilding are unrelated.
   b. are opposites.
   c. exist on a continuum.
   d. are not gendered experiences.

6.   All of the following statements are <u>correct</u> regarding sexism, <u>except</u>: (p. 354)
     a.   sexism, like racism, is used to justify discriminatory treatment.
     b.   evidence of sexism is found in the undervaluation of women's work.
     c.   women may experience discrimination in leisure activities.
     d.   men cannot be the victims of sexist assumptions.

7.   In most hunting and gathering societies: (p. 355)
     a.   a relatively equitable relationship exists because neither sex has the ability to provide all of the food necessary for survival.
     b.   women are not full economic partners with men.
     c.   relationships between women and men tend to be patriarchal in nature.
     d.   menstrual taboos place women in subordinate positions by monthly segregation into menstrual huts.

8.   Gender inequality and male dominance became institutionalized in _____ societies. (p. 357)
     a.   hunting and gathering
     b.   horticultural and pastoral
     c.   agrarian
     d.   industrial

9.   Most common in parts of India, suttee refers to the: (p. 358)
     a.   custom of thwarting the growth of a female's feet.
     b.   seclusion of women, including extreme modesty in apparel.
     c.   surgical procedure performed on young girls as a method of sexual control.
     d.   sacrificial killing of a widow upon the death of her husband.

10.   All of the following statements regarding gender socialization are correct, <u>except</u>: (p. 361)
     a.   virtually all gender roles have changed dramatically in recent years.
     b.   many parents prefer boys to girls because of stereotypical ideas about the relative importance of males and females to the future of the family and society.
     c.   parents tend to treat baby boys more roughly than baby girls.
     d.   parents' choices of toys for their children are not likely to change in the near future.

11. In their investigation of college women, anthropologists Dorothy C. Holland and Margaret A. Eisenhart determined that: (p. 364)
    a. peer groups on college campuses are not organized around gender relations.
    b. the peer system propelled women into a world of romance in which their attractiveness to men counted most.
    c. the women's peers immediately influenced their choices of majors and careers.
    d. peer pressure did not involve appearance norms.

12. Teachers who devote more time, effort, and attention to boys than to girls in schools are an example of: (p. 365)
    a. gender socialization.
    b. gender identity.
    c. gender role differentiation.
    d. gender bias.

13. A comprehensive study of gender bias in schools suggested that girls' self-esteem is undermined in school through all of the following experiences, except: (p. 365)
    a. a relative lack of attention from teachers.
    b. sexual harassment from male peers.
    c. an assumption that girls are better in visual-spatial ability, as compared with verbal ability.
    d. the stereotyping and invisibility of females in textbooks.

14. Which of the following statements is true regarding men and women in higher education? (p. 366)
    a. Men are more likely than women to earn a college degree.
    b. Men tend to have higher grade-point averages in college.
    c. Women typically have higher scores on standardized admissions tests like the Scholastic Assessment Test.
    d. Men constitute the majority of majors in architecture, engineering, computer technology, and physical science.

15. According to the text's discussion of gender and athletics: (p. 366)
    a. women who engage in activities that are assumed to be "masculine" (such as bodybuilding) may either ignore their critics or attempt to redefine the activity or its result as "feminine" or womanly.
    b. some women believe they are more likely to win women's bodybuilding competitions if they "overbuild" their bodies.
    c. being active in sports such as gymnastics makes women less likely to be the victims of anorexia and bulimia.
    d. women bodybuilders have learned that they are less likely to win competitions if they look like a fashion model.

16. According to recent research by sociologist Elizabeth Higginbotham, African American professional women: (p. 370)
    a. find themselves limited in employment in certain sectors of the labor market.
    b. are concentrated in public sector employment.
    c. frequently are employed as public school teachers, welfare workers, librarians, public defenders, and faculty members at public colleges.
    d. all of the above.

17. The belief that wages ought to reflect the worth of a job, not the gender or race of the worker, is referred to as: (p. 373)
    a. the earnings ratio.
    b. pay equity.
    c. non-comparable worth.
    d. the pay gap.

18. According to the functionalist/human capital model: (p. 376)
    a. what women earn is the result of their own choices and the needs of the labor market.
    b. the gendered division of labor in the workplace results from male control of and dominance over women and resources.
    c. women's oppression results from their dual roles as paid and unpaid workers in a capitalist economy.
    d. none of the above.

19. Critics of functionalist and neoclassical economic perspectives have pointed out that these perspectives: (p. 377)
    a. exaggerate the problems inherent in traditional gender roles.
    b. fail to critically assess the structure of society that makes educational and occupational opportunities more available to some than others.
    c. overemphasize factors external to individuals that contribute to the oppression of white women and people of color.
    d. focus on differences between men and women without taking into account the commonalities they share.

20. A women's group fights for better child-care options, women's right to choose an abortion, and the elimination of sex discrimination in the workplace, arguing that the roots of women's oppression lie in women's lack of equal civil rights and educational opportunities. The platform of this group reflects: (p. 378)
    a. black (African American) feminism.
    b. socialist feminism.
    c. radical feminism.
    d. liberal feminism.

TRUE-FALSE QUESTIONS

T   F   1. Many men compare themselves unfavorably to bodybuilders and believe they need to gain weight or muscularity. (p. 347)

T   F   2. Most "sex differences" actually are socially constructed "gender differences." (p. 352)

T   F   3. Gendered belief systems generally do not change over time. (p. 353)

T   F   4. Sexism is the subordination of one sex, usually female, based on the assumed superiority of the other sex. (p. 354)

T   F   5. The earliest known division of labor between women and men is in agrarian societies. (p. 355)

T   F   6. As societies industrialize, the status of women tends to decline further. (p. 360)

T   F   7. The cult of true womanhood increased white women's dependence on men and became a source of discrimination against women of color. (p. 360)

T   F   8. Unlike parents in the United States, parents in many other countries do not prefer boys to girls. (p. 361)

T   F   9. Most studies of gender socialization by parents have been based on the experiences of white, middle-class families. (p. 362)

T   F   10. Many parents are aware of the effect that gender socialization has on their children and make a conscientious effort to provide nonsexist experiences for them. (p. 363)

T   F   11. Female peer groups place more pressure on girls to do "feminine" things than male peer groups place on boys to do "masculine" things. (p. 363)

T   F   12. By young adulthood, men and women no longer receive gender-related messages from peers. (p. 364)

T   F   13. Many teachers use sex segregation as a way to organize students. (p. 366)

T   F   14. Advertising often targets very young girls and suggests that they should learn how to control their weight through the use of some product. (p. 369)

T   F   15. The degree of gender segregation in the professional labor market has declined since the 1970s. (p. 370)

SOCIOLOGY IN OUR TIMES: DIVERSITY ISSUES

1. Think about your own body consciousness -- how a person perceives and feels about his or her body. What do you consider to be your most positive attributes? The most negative? How are your feelings about your body linked to the gender belief system in society?

2. According to sociologist Becky W. Thompson, eating problems exist among women of color, working-class women, lesbians, and some men. Why do many people assume that the primary victims of eating problems are white, middle-class, heterosexual women? Do you think

men of color -- as well as white men -- are affected by "ideal" images of masculinity in the United States?

3.   Do you consider your sex/gender to be an asset, a liability, or neither? Can racism and sexism be used to justify discriminatory treatment of some people in a society?

4.   As a woman or a man, what do you think your life would have been like if you had lived in a hunting and gathering society? A horticultural or pastoral society? An agrarian society? What is the relationship between gender and daily life -- public and private -- in industrial and postindustrial societies?

5.   Why is pay equity, or comparable worth, an important issue for all workers, not just women? (p. 374) Are you planning to enter a male-dominated job? A female-dominated job? If you are a man, are you likely to seek employment in female-dominated fields? If you are a woman, are you likely to seek employment in male-dominated fields?

CHAPTER TEN CROSSWORD PUZZLE

For those who enjoy crossword puzzles, here is a puzzle that contains words and names from Chapter Ten. Working the puzzle will help you in reviewing the chapter. The answers appear on page 201.

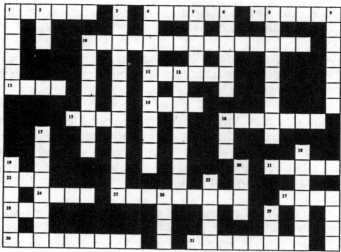

ACROSS

1. Either of the two education professors who identified four types of teacher comments
4. Menstrual _____ place women in a subordinate position
7. The culturally and socially constructed differences between females and males
10. Primary sex _____: the genitalia used in the reproductive process
11. A gender belief _____ includes all of the ideas regarding masculine and feminine attributes that are held to be valid in a society
13. A combination of an x and a y chromosome produces a _____ embryo
14. In other words, the embryo has one of _____
15. _____ elementary school through high school boys play football and other competitive sports
16. The earliest known division of labor between women and men is in ____ and gathering societies
21. Development of secondary sex characteristics produces an _____ awareness of sexuality
22. _____ and all of the following are examples of practices in agrarian societies that contribute to the subordination of women: purdah, footbinding, suttee and genital mutilation
24. Male peer groups place more pressure on a boy to do "masculine" things than female peer groups place on a _____ to do "feminine" things
25. In some pre-industrial societies, women are segregated into _____ huts for the duration of their monthly flow
27. Gender _____: showing favoritism toward one gender over the other
28. Sexual orientation: "straight," _____, lesbian, or bisexual
30. Comparable worth
31. The loss of at least 25% of body weight due to a compulsive fear of becoming fat

DOWN

1. The subordination of one sex, usually female, based on the assumed superiority of the other sex
2. One study found that women college basketball players, on the basketball court "_____ athlete"
3. Rethinking the experiences of African Americans from a feminist perspective
4. A person who believes that he or she was born with the body of the wrong sex
5. The job market reinforces the image of female beauty as childlike and thin through ____ and covert discrimination against women who do not fit the image
6. We objectify people when we judge them on the basis of their status in a _____-laden category such as "fat slob"
8. First name of sociologist who noted that African American professional women find themselves limited to employment in certain sectors of the labor market
9. According to _____ feminists, male domination causes all forms of human oppression, including racism and classism
10. Throughout life, men and women receive different _____ messages about body image, food, and eating
12. _____ feminists suggest that women's oppression results from their dual role as paid and unpaid workers in a capitalist economy
16. In horticultural societies, people are able to grow their own food because of hand tools, such as the digging stick and the _____
17. The marriage of one man to multiple wives
18. With _____, a person binges by consuming large quantities of food and then purges the food by induced vomiting, etc.
19. The disparity between women's and men's earnings
20. _____ of domesticity
23. _____ capital is acquired by education and job training
26. Although in days gone by, it was considered a sign of wealth to be "pleasingly plump," today many people wish to be _____
29. The biological and anatomical differences between females and males

198

ANSWERS TO PRACTICE TEST, CHAPTER 10

Answers to Multiple Choice Questions

1. d   In responding to the question, "Why do women and men feel differently about their bodies?" the text points out that: cultural differences in appearance norms may explain women's greater concern; in the United States, the image of female beauty as childlike and thin is continually flaunted by the advertising industry; and women of all racial-ethnic groups, classes, and sexual orientations regard their weight as a crucial index of their acceptability to others. Thus, "all of the above" is the correct answer. (p. 348)

2. b   Sociologists use the term sex to refer to the biological attributes of men and women; gender is used to refer to the distinctive qualities of men and men that are culturally created. (p. 349)

3. a   A(n) hermaphrodite is a person in whom sexual differentiation is ambiguous or incomplete. (p. 351)

4. b   A person's perception of the self as female or male is referred to as one's gender identity. (p. 352)

5. c   According to historian Susan Bordo, the anorexic body and the muscled body exist on a continuum. (p. 354)

6. d   All of the following statements are correct regarding sexism, except: men cannot be the victims of sexist assumptions. (p. 354)

7. a   In most hunting and gathering societies a relatively equitable relationship exists because neither sex has the ability to provide all of the food necessary for survival. (p. 355)

8. c   Gender inequality and male dominance became institutionalized in agrarian societies. (p. 357)

9. d   Most common in parts of India, suttee refers to the sacrificial killing of a widow upon the death of her husband. (p. 358)

10. a   All of the following statements regarding gender socialization are correct, except: virtually all gender roles have changed dramatically in recent years. (p. 361)

11. b   In their investigation of college women, anthropologists Dorothy C. Holland and Margaret A. Eisenhart determined that the peer system propelled women into a world of romance in which their attractiveness to men counted most. (p. 364)

12. d   Teachers who devote more time, effort, and attention to boys than to girls in schools are an example of gender bias. (p. 365)

13. c   A comprehensive study of gender bias in schools suggested that girls' self-esteem is undermined in school through all of the following experiences, except: an assumption that girls are

better in visual-spatial ability, as compared with verbal ability. (p. 365)

14.   c   Which of the following statements is <u>true</u> regarding men and women in higher education? Women typically have higher scores on standardized admissions tests like the Scholastic Assessment Test. (p. 366)

15.   a   According to the text's discussion of gender and athletics women who engage in activities that are assumed to be "masculine" (such as bodybuilding) may either ignore their critics or attempt to redefine the activity or its result as "feminine" or womanly. (p. 366)

16.   d   According to recent research by sociologist Elizabeth Higginbotham, African American professional women find themselves limited in employment in certain sectors of the labor market; are concentrated in public sector employment; and are frequently employed as public school teachers, welfare workers, librarians, public defenders, and faculty members at public colleges. Thus, "all of the above" is the correct answer. (p. 370)

17.   b   The belief that wages ought to reflect the worth of a job, not the gender or race of the worker, is referred to as pay equity. (p. 373)

18.   a   According to the functionalist/human capital model, what women earn is the result of their own choices and needs of the labor market. (p. 376)

19.   b   Critics of functionalist and neoclassical economic perspectives have pointed out that these perspectives fail to critically assess the structure of society that makes educational and occupational opportunities more available to some than others. (p. 377)

20.   d   A women's group fights for better child-care options, women's right to choose an abortion, and the elimination of sex discrimination in the workplace, arguing that the roots of women's oppression lie in women's lack of equal civil rights and educational opportunities. The platform of this group reflects liberal feminism. (p. 378)

Answers to True-False Questions

1.   True (p. 347)
2.   True (p. 352)
3.   False -- Gendered belief systems <u>do</u> change over time as gender roles change. Sometimes these changes are relatively slow, however, because popular stereotypes and existing cultural norms serve to reinforce gendered institutions in society. (p. 353)

4. True (p. 354)
5. False -- The earliest known division of labor between women and men is in <u>hunting and gathering</u> societies. (p. 355)
6. True (p. 360)
7. True (p. 360)
8. False -- Across cultures, parents tend to prefer boys to girls. This is especially true in countries where the number of children that parents can have is limited by law or economic conditions. (p. 361)
9. True (p. 362)
10. True (p. 363)
11. False -- <u>Male</u> peer groups place more pressure on <u>boys</u> to do "<u>masculine</u>" things than <u>female</u> peer groups place on <u>girls</u> to do "<u>feminine</u>" things. (p. 363)
12. False -- As young adults, men and women still receive many gender-related messages from peers. (p. 364)
13. True (p. 366)
14. True (p. 369)
15. True (p. 370)

ANSWER TO CHAPTER TEN CROSSWORD PUZZLE

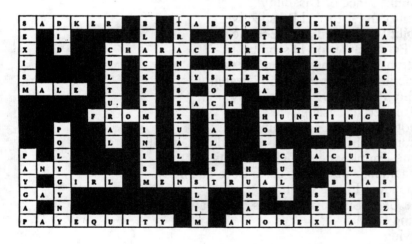

CHAPTER 11
AGING AND DISABILITY

BRIEF CHAPTER OUTLINE
An Overview of Aging and Disability
The Social Significance of Age
        Trends in Aging
        Age in Historical Perspective
        Age in Contemporary Society
Inequalities Related to Aging
        Ageism
        Wealth, Poverty, and Aging
        Elder Abuse
Sociological Perspectives on Aging
        Functionalist Perspective on Aging
        Interactionist Perspective on Aging
        Conflict Perspective on Aging
The Social Significance of Disability
        Disability in Historical Perspective
        Disability in Contemporary Society
Sociological Perspectives on Disability
        Functionalist Perspective on Disability
        Interactionist Perspective on Disability
        Conflict Perspective on Disability
Inequalities Related to Disability
        Stereotypes Based on Disability
        Prejudice and Discrimination Based on Disability
        Income, Employment, and Disability
Living Arrangements and Long-Term Care Facilities
        Independent Living Centers
        Home Care, Support Services, and Day Care
        Nursing Homes
Death and Dying
Aging and Disability in the Twenty-First Century

CHAPTER SUMMARY
Aging is the physical, psychological, and social processes associated with
growing older. In the United States, the proportion of people age 65 and
older is increasing, while the proportion of young people is decreasing. Age
defines what is appropriate or expected of us. In preindustrial societies,
people of all ages are expected to help with the work and the contributions of
older people are valued. In industrialized societies, however, older people

often are expected to retire so that younger people may take their place. Age differentiation in the United States produces age stratification: the inequalities, differences, segregation, or conflict between age groups. **Ageism** -- prejudice and discrimination against people on the basis of age -- is reinforced by stereotypes based on one-sided and exaggerated images of older people. Elder abuse includes physical abuse, psychological abuse, financial exploitation, and medical abuse or neglect of people age 65 or older. Functionalist explanations of aging -- such as **disengagement theory** -- focus on how older persons adjust to their changing roles in society. **Activity theory** -- an interactionist perspective -- states that people change in late middle age and find substitutes for previous statuses, roles, and activities. Conflict theorists link the loss of status and power experienced by many older persons to their lack of ability to produce and maintain wealth in a capitalist economy.

While the medical field tends to define a disability in terms of physical impairment, it may be more correct to define it as a physical or health condition that stigmatizes or causes discrimination. Many persons with a disability are kept out of the mainstream of U.S. society through practices such as assignment to special schools and restricted job opportunities. A functionalist perspective on disabilities emphasizes that a functioning social system requires all people to perform their appropriate social roles. Persons who are sick or disabled cannot perform their designated roles and are thus confined to a sick role. According to the interactionist perspective, some people avoid persons with disabilities and treat them as if they were deviant. According to the conflict perspective, persons in power create policies and barriers that keep people with disabilities in subservient positions where they are subjected to discrimination and employment bias. In contrast, the independent living movement assumes persons with a disability should be able to live and work independently. In industrial societies, death has been removed from everyday life and is often regarded as unnatural.

LEARNING OBJECTIVES
After reading Chapter 11, you should be able to:
1. Distinguish between aging and disability and explain why older persons and persons with a disability share some important concerns.

2. Differentiate between chronological age and functional age and note the social significance of each.

3.   Trace recent trends in aging and explain how life expectancy has changed in the United States during the twentieth century.

4.   Compare historical and contemporary perspectives on age.

5.   Trace the process of aging through the life course, noting the social consequences of age at each stage.

6.   Discuss ageism and describe the negative stereotypes associated with older persons.

7.   Describe the relationship between aging and class, age, race/ethnicity, and gender.

8.   Distinguish between functionalist, interactionist, and conflict perspectives on aging.

9.   Compare historical and contemporary perspectives on disability.

10. Distinguish between functionalist, interactionist, and conflict perspectives on disability.

11. Discuss inequalities related to disability and describe the impact of stereotypes, prejudice, and discrimination on persons with disabilities.

KEY TERMS (defined at page number shown and in glossary)

activity theory 401
ageism 396
age stratification 393
aging 388
chronological age 389
cohort 391
disability 388

disengagement theory 401
elder abuse 400
entitlements 399
functional age 391
hospice 417
life expectancy 391
social gerontology 393

KEY PEOPLE (identified at page number shown)

Elaine C. Cumming and
  William E. Henry 401
Anne Fausto-Sterling 394
Eliot Freidson 409
Melissa A. Hardy and
  Lawrence E. Hazelrigg 399
Madonna Harrington
  Meyer 403, 415

Elisabeth Kubler-Ross 417
William C. Levin 397
Patricia Moore 397
Margaret Nosek 410
Meira Weiss 408

CHAPTER OUTLINE

I.   AN OVERVIEW OF AGING AND DISABILITY

    A.   **Aging** is the physical, psychological, and social processes associated with growing older; a **disability** is a physical or health condition that stigmatizes or causes discrimination.

    B.   Older people and persons with a disability may be targets of prejudice and discrimination; they may need assistance from others and support from society; and they have used similar methods to gain dignity and autonomy.

II.   THE SOCIAL SIGNIFICANCE OF AGE

    A.   **Chronological age** is a person's age based on date of birth; **functional age** is observable physical attributes (physical appearance, mobility, etc.) that are used to assign people to age categories.

    B.   There are significant differences in **life expectancy** -- the average length of time a group of individuals of the same age will live -- based on racial-ethnic and sex differences.

    C.   **Social gerontology** is the study of the social (non-physical) aspects of aging.

    D.   While young persons historically were considered "little adults" and were expected to do adult work, today the skills necessary for many roles are more complex and the number of unskilled positions is more limited.

    E.   **Age stratification** refers to the inequalities, differences, segregation, or conflict between age groups.

III. INEQUALITIES RELATED TO AGING
  A. **Ageism**-- prejudice and discrimination against people on the basis of age, particularly when they are older persons -- is reinforced by stereotypes of older persons as cranky, sickly, and lacking in social value.
  B If we compare wealth (all economic resources of value) with income (available money or its purchasing power), we find that older people tend to have more wealth but less income than younger people; however, race/ethnicity and gender continue to be differentiating factors.
  C. Ninety percent of all retired people in the U.S. draw Social Security benefits, which are a form of **entitlements** -- benefits paid by the government. Another entitlement is Medicare, a nationwide health care program for persons age 65 and over who participate in Social Security or "buy into" the program.
  D. **Elder abuse** refers to physical abuse, psychological abuse, financial exploitation, and medical abuse or neglect of people age 65 or over.
IV. SOCIOLOGICAL PERSPECTIVES ON AGING
  A. Functionalists examine how older persons adjust to their changing roles. According to **disengagement theory**, older persons make a normal and healthy adjustment to aging when they detach themselves from their social roles and prepare for their eventual death.
  B. **Activity theory** -- an interactionist perspective on aging -- states that people tend to shift gears in later life and find substitutes for previous statuses, roles, and activities.
  C. Conflict theorists note that, as people grow older, their power tends to diminish unless they are able to maintain their wealth. Consequently, those who have been disadvantaged in their younger years become even more so in late adulthood.
V. THE SOCIAL SIGNIFICANCE OF DISABILITY
  A. Disability has existed in all societies throughout human history. How a particular society has dealt with disability differed on the basis of culture, values, and technology.
  B. An estimated 48 million persons in the U.S. have one or more physical or mental disabilities, and the number is increasing as medical advances make it possible for those who would have died from an accident or illness to survive (but with an impairment) and as life expectancies increase.
  C. Environment, lifestyle, and working conditions may contribute to disability.
  D. Many disability rights advocates argue that persons with a disability are kept out of the mainstream of society -- e.g.,

denied equal educational opportunities by being consigned to special classes or schools.

VI.   SOCIOLOGICAL PERSPECTIVES ON DISABILITY
- A.   Functionalist Talcott Parsons focused on how people who are ill (or disabled) fill the **sick role**. This is a "medical model" of disability: persons with disabilities become chronic patients.
- B.   Interactionist perspectives examine how people are labeled as a result of disability. Eliot Freidson determined that the particular label results from (1) the person's degree of responsibility for the impairment, (2) the apparent seriousness of the condition, and (3) the perceived legitimacy of the condition.
- C.   From a conflict perspective, persons with disabilities are a subordinate group in conflict with persons in positions of power in the government, the health care industry, and the rehabilitation business. When people with disabilities are defined as a social problem and public funds are spent to purchase goods and services for them, rehabilitation becomes big business.

VII.   INEQUALITIES RELATED TO DISABILITY
- A.   Stereotypes of persons with a disability fall into two categories: (1) deformed individuals who also may be horrible deviants; or (2) persons who are to be pitied. Even apparently positive stereotypes ("supercrips") can be harmful to persons with a disability.
- B.   Prejudice against persons with disabilities may result in either subtle or overt discrimination.

VIII.   LIVING ARRANGEMENTS; LONG-TERM CARE FACILITIES
- A.   The independent-living movement is based on the concept that persons with a disability should have the chance to live like other people and work independently.
- B.   Many persons with a disability, as well as frail, older persons, live alone or in a family setting with care provided informally by family members.
- C.   Nursing homes are the most restrictive environment for older persons and persons of all ages with a disability.

IX.   DEATH AND DYING
- A.   In contemporary industrial societies, death is looked on as unnatural because it has been removed from everyday life: most deaths occur among older persons and in institutionalized settings.
- B.   Elizabeth Kubler-Ross proposed that people cope with dying in five stages:

1.   denial ("Not me!");
2.   anger ("Why me?");
3.   bargaining ("Yes me, but....");
4.   depression and sense of loss; and
5.   acceptance.
    C.   A **hospice** is a homelike facility that provides supportive care for patients with terminal illnesses.

X.   AGING AND DISABILITY IN THE TWENTY-FIRST CENTURY
    A.   By the year 2050, there will be approximately 80 million persons age 65 and over, as compared with 33 million today. More people will survive to age 85 -- or even 95 and more.
    B.   A 1994 report warns that this increase (coupled with a decreasing birth rate) may result in entitlements consuming nearly all federal tax revenues by the year 2012.

ANALYZING AND UNDERSTANDING THE BOXES

After reading the chapter and studying the outline, re-read the four boxes and write down key points and possible questions for class discussion.

Sociology and Everyday Life -- "How Much Do You Know About Aging, Disability, and Empowerment?"

Key Points:

Discussion Questions:

1.

2.

3.

Sociology and Law -- "The Americans with Disabilities Act (ADA) and Empowerment"

Key Points:

Discussion Questions:

1.

2.

3.

Sociology and Media -- "Stereotypes, Disability, and Fund-Raising"

Key Points:

Discussion Questions:

1.

2.

3.

Sociology in Global Perspective -- "The Politics of Disability in China"

Key Points:

Discussion Questions:

1.

2.

3.

PRACTICE TEST

MULTIPLE CHOICE QUESTIONS

Select the response that best answers the question or completes the statement:

1. All of the following statements regarding aging and disability are correct, except: (p. 388)
    a. Aging is the physical, psychological, and social processes associated with growing older.
    b. Disability is a physical or health condition that stigmatizes or causes discrimination.
    c. Many disabilities are not visible to others.
    d. As people age, they inevitably become disabled.

2. _____ age refers to a person's age based on date of birth; by contrast, _____ age refers to observable individual attributes such as physical appearance, mobility, strength, coordination, and mental capacity that are used to assign people to age categories. (pp. 390-391)
    a. Obvious -- subjective
    b. Subjective -- obvious
    c. Chronological -- functional
    d. Functional -- chronological

3. The median age of the U.S. population is: (p. 391)
    a. increasing due to an increase in life expectancy combined with a decrease in birth rates.
    b. decreasing due to an increase in birth rates.
    c. roughly the same as it has been for most of the twentieth century due to a stabilization of both birth and death rates.
    d. unaffected by the aging of the Baby Boomers.

4. Two hundred years ago, the age spectrum was divided into_____. (p. 393)
    a. babyhood and adulthood.
    b. babyhood, a very short childhood, and adulthood.
    c. babyhood, childhood, adulthood, and old age.
    d. infancy, childhood, adolescence, adulthood, and old age.

5.	According to the text, middle adulthood represents the time when many people: (p. 394)
	a.	are strapped for cash because their children are in college.
	b.	may have grandchildren who give them another tie to the future.
	c.	have a decline in income as they approach their retirement years.
	d.	grow discontented with their spouse of many years.

6.	Which of the following statements is <u>correct</u> regarding Alzheimer's disease? (p. 395)
	a.	Most older people suffer from Alzheimer's disease.
	b.	Most people with Alzheimer's disease have an extremely short life expectancy.
	c.	About 55 percent of all organic mental disorders in the older population is caused by Alzheimer's disease.
	d.	In 1995, researchers found a possible cure for Alzheimer's disease.

7.	According to Erik Erikson, older people must resolve a tension of _____ versus _____ in their lives. (p. 396)
	a.	integrity -- despair
	b.	activity -- disengagement
	c.	stability -- change
	d.	autonomy -- entitlements

8.	Ageism is: (p. 396)
	a.	prejudice and discrimination against people on the basis of age, particularly when they are younger persons.
	b.	prejudice and discrimination against people on the basis of age, particularly when they are older persons.
	c.	gradually being overcome by positive depictions of older persons in the media.
	d.	experienced equally by both women and men.

9.	When Patricia Moore, at age 27, disguised herself as an 85-year-old woman and went to a grocery store, she learned that: (p. 397)
	a.	people permitted her to go to the head of the checkout line because of her age.
	b.	grocery store clerks treated her as a preferred customer.
	c.	people tried to have as little to do with her as possible.
	d.	other people's reactions to her changed when she played the role of an older person.

10. According to the text, if we compare older people with younger people, we will find that older people as a category tend to have more _____ but less _____ than younger people. (p. 398)
    a. morals -- stamina
    b. life insurance -- health insurance
    c. wealth -- income
    d. entitlements -- wealth

11. In a recent study, gerontologists Melissa H. Hardy and Lawrence E. Hazelrigg found that: (p. 399)
    a. gender is more directly related to poverty in older persons than is race/ethnicity, educational background, or occupational status.
    b. race/ethnicity is more directly related to poverty in older persons than is gender, educational background, or occupational status.
    c. educational background is more directly related to poverty in older persons than is gender, race/ethnicity, or occupational status.
    d. occupational status is more directly related to poverty in older persons than is gender, race/ethnicity, or educational background.

12. All of the following statements are <u>correct</u> regarding elder abuse, <u>except</u>: (p. 400)
    a. abuse and neglect of older persons has received increasing public attention in recent years.
    b. financial exploitation is a form of elder abuse.
    c. more than 1.5 million older people in the United States are victims of elder abuse.
    d. nursing home personnel are the most frequent abusers of older persons.

13. Functionalist perspectives on aging focus on: (p. 401)
    a. how older persons adjust to their changing roles in society.
    b. the connection between personal satisfaction in a person's later years and a high level of activity.
    c. reasons why aging is especially problematic in contemporary capitalist societies.
    d. how people are consigned to the sick role.

14. _____ theory states that people tend to shift gears in late middle age and find substitutes for previous statuses, roles, and activities. (p. 401)
    a. Continuity
    b. Disengagement
    c. Activity
    d. Engagement

15. According to sociologists, disability is best viewed as: (p. 404)
    a. residing primarily in the individual who exhibits the disability.
    b. residing primarily in social attitudes and in social and physical environments.
    c. an examination of how persons with disabilities can adapt to the mainstream environment.
    d. an organically based impairment.

16. In contemporary industrial societies, disability often can be attributed to: (p. 404)
    a. epidemics related to poor sanitation and overcrowding.
    b. urban density and poverty.
    c. environment, lifestyle, and working conditions.
    d. employment in high stress jobs in the primary tier of the labor market.

17. According to Talcott Parsons, persons who assume the sick role: (p. 409)
    a. are responsible for their condition.
    b. must try to get well.
    c. cannot be made exempt from their normal roles and obligations or they may unnecessarily prolong their "illness."
    d. typically are mentally ill.

18. Disabilities are big business, according to _____. (p. 409)
    a. disability rights advocates.
    b. interactionist theorists.
    c. functionalist theorists.
    d. conflict theorists.

19. According to sociologists, treating persons with disabilities as asexual is an example of _____ discrimination. (p. 410)
    a. subtle
    b. overt
    c. inhumane
    d. institutional

20. In her analysis of the financing of nursing home care, sociologist Madonna Harrington Meyer concluded that the United States has a long-term care system that is based on and perpetuates: (p. 415)
    a.   racial inequality.
    b.   class inequality.
    c.   gender inequality.
    d.   all of the above.

TRUE-FALSE QUESTIONS

T   F   1.   Persons with certain types of disabilities are perceived as much older than their actual age because of physical or psychological frailties associated with their disability.

T   F   2.   Due to the ongoing aging of the U.S. population and advances in medical technology, many of us can expect to live for a number of years with illness and disability. (p. 389)

T   F   3.   About 4 percent of the U.S. population was over age 65 in 1993. (p. 391)

T   F   4.   Social gerontology is the study of the non-physical aspects of aging. (p. 393)

T   F   5.   In contemporary societies, adolescence is treated as a continuation of childhood. (p. 393)

T   F   6.   The cosmetics industry helps perpetuate the myth that age reduces the "sexual value" of women but increases it for men. (p. 396)

T   F   7.   Few of the wealthiest people in the United States are over 65 years of age. (p. 398)

T   F   8.   Age, race/ethnicity, gender, and poverty are interrelated. (p. 398)

T   F   9.   According to analysts, Social Security keeps more white men out of poverty than it does white women and people of color. (p. 400)

T   F   10.   Disability is a relatively recent occurrence because of longer life spans. (p. 404)

T	F	11.	The Americans with Disabilities Act (ADA) was designed to end discrimination in all areas of life for persons with a disability. (p. 405)
T	F	12.	In a study of children with disabilities, sociologist Meira Weiss found that parents automatically bond with infants born with visible disabilities. (p. 408)
T	F	13.	Functionalists have suggested that persons with a disability experience role ambiguity because many people equate disability with deviance. (p. 409)
T	F	14.	"Supercrip" stereotypes may be harmful to persons with a disability. (p. 410)
T	F	15.	About two-thirds of working-age persons with a disability in the United States are unemployed. (p. 413)

SOCIOLOGY IN OUR TIMES: DIVERSITY ISSUES

1. Do you have older relatives and acquaintances? If so, have they experienced some of the problems discussed in this chapter? Can you think of macrolevel changes that might enhance their daily lives?

2. As you were growing up, did you spend most of your time with persons of similar ages to your own? If so, do you think age segregation had any affect on your outlook regarding age?

3. Why are stereotypes about aging especially damaging to women? Do you think of older women differently from older men? Why or why not?

4. Can you find current examples in the media that highlight the intertwining of age, race/ethnicity, gender, and poverty? For example, news stories often show an older African American woman as the typical "poverty-ridden" individual. Is this an accurate depiction? Why or why not?

5. According to the text, "disability knows no socioeconomic boundaries." If this statement is true, why do persons with lower incomes have higher rates of disability than persons with higher incomes?

6. What measures have been taken by your college or university to "mainstream" students with a disability? Do you think these measures have been adequate? Does more need to be done? If you are a student with a disability, do you believe that you have equal access to educational opportunities at your institution? Why or why not?

CHAPTER ELEVEN CROSSWORD PUZZLE

For those who enjoy crossword puzzles, here is a puzzle that contains words and names from Chapter Eleven. Working the puzzle will help you in reviewing the chapter. The answers appear on page 220.

ACROSS

1. _____ theory: older persons make a normal and healthy adjustment to aging when they detach themselves from their social roles

7. A homelike facility that provides supportive [17 down] for patients with terminal illnesses

11. Activity theory: people tend to _____ gears in late middle age and find substitutes for previous statuses, roles and activities

12. A certain type of benefit payment paid by the government

13. Chronological age: a person's age based on his or ___ age

15. Men may have a ___ life crisis in which they assess what they have accomplished

16. _____ adulthood begins at age 65

17. While _____ demand does decrease for older people, their nutritional needs remain the same

21. The _____ living movement encourages innovative living arrangements for persons with a disability

23. According to Talcott Parsons, when individuals are ill, they are consigned to the sick ____

24. Madonna Harrington _____ described problems inherent in financing nursing home care

25. Adolescence roughly spans the teenage _____

27. The physical, psychological, and social processes associated with growing older

28. She conducted research on sexuality issues faced by women with physical disabilities

30. Age, race, _____, and poverty are closely intertwined

31. _____ gerontology studies the nonphysical aspects of aging

32. _____ abuse: psychological abuse, financial exploitation, and medical abuse or neglect of people age 65 or over

DOWN

1. A physical or health condition that stigmatizes or causes discrimination

2. Conflict theorists assert that, to minimize demand for governmental services for the elderly, these services are made punitive and a _____ is attached to them

3. Elaine C. Cumming and William E. Henry _____ that [1 across] can be functional for both the individual and society

4. Prejudice and discrimination against people on the basis of age, particularly when [6 down] are older persons

5. Sociologist _____ Weiss challenged the assumption that parents automatically bond with infants with a visible disability

6. See 4 down

8. Persons with Alzheimer's eventually lose all sense of their ___ identity

9. A group of people within a specified period of time

10. Role _____: the number and type of positions open to persons of each age level within the society

14. Often asked question: "How ___ are you?"

16. According to Eliot Freidson, the _____ disabled people receive results from three factors, one of which is the apparent seriousness of their condition

17. See 7 across

18. For persons with _____ illness and disability, life expectancy may take on a different meaning

19. Until the 20th century, poor nutrition, infectious diseases, accidents, and natural disasters took their toll on ___ and women of all ages

20. Elizabeth Kubler-Ross proposed five _____ of reaction upon learning you are dying

22. In the past, explanations for _____ and dying were rooted in custom and religious beliefs

24. Patricia _____, at age 27, disguised herself as an 85-year-old woman to see how people reacted on the basis of appearance.

26. Many disabilities are ___ visible to other people

29. A person with a terminal illness is near the ___ of their life

217

ANSWERS TO PRACTICE TEST, CHAPTER 11

Answers to Multiple Choice Questions

1.   d   All of the following statements regarding aging and disability are correct, <u>except</u>: as people age, they inevitably become disabled. (p. 388)

2.   c   Chronological age refers to a person's age based on date of birth; by contrast, functional age refers to observable individual attributes such as physical appearance, mobility, strength, coordination, and mental capacity that are used to assign people to age categories. (pp. 390-391)

3.   a   The median age of the U.S. population is increasing due to an increase in life expectancy combined with a decrease in birth rates. (p. 391)

4.   b   Two hundred years ago, the age spectrum was divided into babyhood, a very short childhood, and adulthood. (p. 393)

5.   b   According to the text, middle adulthood represents the time when many people may have grandchildren who give them another tie to the future. (p. 394)

6.   c   Which of the following statements is <u>correct</u> regarding Alzheimer's disease? About 55 percent of all organic mental disorders in the older population is caused by Alzheimer's disease. (p. 395)

7.   a   According to Erik Erikson, older people must resolve a tension of integrity versus despair in their lives. (p. 396)

8.   b   Ageism is prejudice and discrimination against people on the basis of age, particularly when they are older persons. (p. 396)

9.   d   When Patricia Moore, at age 27, disguised herself as an 85-year-old woman and went to a grocery store, she learned that other people's reactions to her changed when she played the role of an older person. (p. 397)

10.   c   According to the text, if we compare older people with younger people, we will find that older people as a category tend to have more wealth but less income than younger people. (p. 398)

11.   a   In a recent study, gerontologists Melissa H. Hardy and Lawrence E. Hazelrigg found that gender is more directly related to poverty in older persons than is race/ethnicity, educational background, or occupational status. (p. 399)

12.   d   All of the following statements are <u>correct</u> regarding elder abuse, <u>except</u>: nursing home personnel are the most frequent abusers of older persons. Instead, sons, followed by daughters, are the most frequent abusers of older persons. (p. 400)

13. a Functionalist perspectives on aging focus on how older persons adjust to their changing roles in society. (p. 401)
14. c Activity theory states that people tend to shift gears in late middle age and find substitutes for previous statuses, roles, and activities. (p. 401)
15. b According to sociologists, disability is best viewed as residing primarily in social attitudes and in social and physical environments. (p. 404)
16. c In contemporary industrial societies, disability often can be attributed to environment, lifestyle, and working conditions. (p. 404)
17. b According to Talcott Parsons, persons who assume the sick role must try to get well. (p. 409)
18. d Disabilities are big business, according to conflict theorists. (p. 409)
19. a According to sociologists, treating persons with disabilities as asexual is an example of subtle discrimination. (p. 410)
20. d In her analysis of the financing of nursing home care, sociologist Madonna Harrington Meyer concluded that the United States has a long-term care system that is based on and perpetuates racial, class, and gender inequalities, thus making the best response "all of the above." (p. 415)

Answers to True-False Questions

1. True (p. 388)
2. True (p. 389)
3. False -- About 13 percent of the U.S. population was over age 65 in 1993. (p. 391)
4. True (p. 393)
5. False -- In contemporary societies, adolescence is a period in which the individual is neither treated as a child nor afforded full status as an adult. (p. 393)
6. True (p. 396)
7. False -- Many of the wealthiest people are over 65 years of age; however, almost 13 percent of all people over 65 live in poverty. (p. 398)
8. True (p. 398)
9. True (p. 400)
10. False -- Disability has existed in all societies throughout human history. (p. 404)
11. True (p. 405)
12. False -- In a study of children with disabilities, sociologist Meira Weiss found that parents <u>did not</u> automatically bond with infants born

with visible disabilities. She found that an infant's appearance may determine how parents will view the child. (p. 408)

13. False -- <u>Interactionists</u> have suggested that persons with a disability experience role ambiguity because many people equate disability with deviance. (p. 409)
14. True (p. 410)
15. True (p. 413)

ANSWER TO CHAPTER ELEVEN CROSSWORD PUZZLE

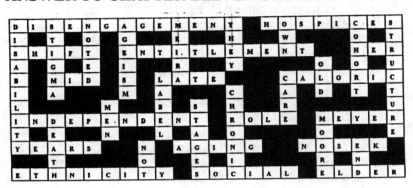

CHAPTER 12
THE ECONOMY AND WORK

BRIEF CHAPTER OUTLINE
The Economy
    The Sociology of Economic Life
    Historical Changes in Economic Systems
Contemporary Economic Systems
    Capitalism
    Socialism
    Mixed Economies
Perspectives on Economy and Work in the United States
    Functionalist Perspective
    Conflict Perspective
The Social Organization of Work
    Job Satisfaction and Alienation
    Occupations
    Professions
    Managers and the Managed
    The Lower Tier of the Service Sector and Marginal Jobs
    Contingent Work
Unemployment
Worker Resistance and Activism
    Labor Unions
    Absenteeism, Sabotage, and Resistance
The Global Economy in the Twenty-first Century
    The U.S. Economy
    Global Economic Interdependence and Competition

CHAPTER SUMMARY
The **economy** is the social institution that ensures the maintenance of society through the production, distribution, and consumption of goods and services. Preindustrial societies are characterized by **primary sector production** in which workers extract raw materials and natural resources from the environment and use them without much processing. Industrial societies engage in **secondary sector production** which is based on the processing of raw materials (from the primary sector) into finished goods. The economic base in postindustrial societies shifts to **tertiary sector production** -- the provision of services rather than goods. As an ideal type, **capitalism** has four distinctive features: private ownership of the means of production, pursuit of personal profit, competition, and lack of government intervention. **Socialism** is characterized by public ownership of the means of production, the pursuit

of collective goals, and centralized decision making. In a **mixed economy**, elements of a capitalist, market economy are combined with elements of a command, socialist economy. Mixed economies are often referred to as **democratic socialism** -- an economic and political system that combines private ownership of some of the means of production, governmental distribution of some essential goods and services, and free elections. For many people, jobs and professions are key sources of identity. **Occupations** are categories of jobs that involve similar activities at different work sites. **Professions** are high-status, knowledge-based occupations characterized by abstract, specialized knowledge, autonomy, authority over clients and subordinate occupational groups, and a degree of altruism. Those in managerial occupations typically are responsible for workers, physical plants, equipment, and the finances of a bureaucratic organization. **Marginal jobs** are those that do not comply with the following employment norms: job content should be legal, jobs should be covered by government regulations, jobs should be relatively permanent, and jobs should provide adequate hours and pay in order to make a living. **Contingent work** is part-time work, temporary work, and subcontracted work that offers advantages to employers but may be detrimental to workers. **Unemployment** remains a problem for many workers. As we approach the twenty-first century, the gap between the rich and the poor is growing.

LEARNING OBJECTIVES
After reading Chapter 12, you should be able to:

1.   Describe the purpose of the economy and distinguish between sociological perspectives on the economy and the study of economics.

2.   Trace the major historical changes that have occurred in economic systems and note the most prevalent form of production found in each.

3.   Describe the four distinctive features of "ideal" capitalism and explain why pure capitalism does not exist.

4.   Discuss socialism and describe its major characteristics.

5.   Compare and contrast capitalism, socialism, and mixed economies.

6.   Distinguish between functionalist and conflict perspectives on the economy and work.

7.   Describe job satisfaction and alienation, and explain the impact of each on workers.

8.   Distinguish between the primary and secondary labor markets and describe the types of jobs found in each.

9.   Discuss the major characteristics of professions and describe the process of deprofessionalization.

10.   Compare scientific management (Taylorism) with mass production through automation (Fordism), noting the strengths and weaknesses of each.

11.   Identify the occupational categories considered to be marginal jobs and explain why they are considered to be marginal.

12.   Discuss contingent work and identify the role subcontracting often plays in contingent work.

13.   Distinguish between the various types of unemployment and explain why the unemployment rate may not be a true reflection of unemployment in the United States.

14. Trace the development of labor unions and describe some of the means by which workers resist working conditions they consider to be oppressive.

KEY TERMS (defined at page number shown and in glossary)

capitalism 432
conglomerates 438
contingent work 453
corporations 433
democratic socialism 442
economy 426
interlocking corporate
   directorates 438
labor union 432
marginal jobs 450
mixed economy 442
multinational corporations 433

occupations 445
oligopoly 437
primary labor market 445
primary sector production 429
professions 446
secondary labor market 445
secondary sector production 429
shared monopoly 437
socialism 440
subcontracting 453
tertiary sector production 431
unemployment rate 454

KEY PEOPLE (identified at page number shown)

Daniel Bell 431
Esther Ngan-Ling Chow 457
Henry Ford 449
Karen J. Hossfeld 453
Karl Marx 444

George Ritzer 431, 449
Denise Segura 450
Adam Smith 438
Frederick Winslow Taylor 448

CHAPTER OUTLINE
I. THE ECONOMY
   A.   The **economy** is the social institution that ensures the maintenance of society through the production, distribution, and consumption of goods and services.
   B.   While economists attempt to explain how the limited resources and efforts of a society are allocated among competing ends, sociologists focus on interconnections between the economy, other social institutions, and the social organization of work.
   C.   Historical Changes in Economic Systems
        1.   In preindustrial societies, most workers engage in **primary sector production** -- the extraction of raw materials and natural resources from the environment.
        2.   Industrialization brings sweeping changes to the economy as new forms of energy and machine technology proliferate as the primary means of

producing goods and most workers engage in **secondary sector production** -- processing raw materials into finished goods.

3. A postindustrial economy is based on **tertiary sector production** -- the provision of services (such as food service, transportation, communication, education, and entertainment) rather than goods.

II. CONTEMPORARY ECONOMIC SYSTEMS

A. **Capitalism** is an economic system characterized by private ownership of the means of production, from which personal profits can be derived through market competition and without government intervention.

B. "Ideal" capitalism has four distinctive features:
1. Private ownership of the means of production
2. Pursuit of personal profit
3. Competition; and
4. Lack of government intervention.

C. However, ideal capitalism does not exist in the U.S. for a number of reasons, including the presence of:
1. **Oligopolies** -- where several companies control an entire industry.
2. **Shared monopolies** -- where four or fewer companies supply 50 percent or more of a particular market.
3. Mergers and acquisitions across industries create **conglomerates** -- combinations of businesses in different commercial areas, all owned by one holding company.
4. Competition also is reduced by **interlocking corporate directorates** -- members of the board of directors of one corporation who also sit on the board(s) of other corporations.
5. Government intervention often occurs in the form of regulations after some individuals and companies in pursuit of profits have run roughshod over weaker competitors; however, much "government intervention" has been in the form of aid to business (tax credits, loan guarantees, etc.).

D. **Socialism** is an economic system characterized by public ownership of the means of production, the pursuit of collective goals, and centralized decision making.

E. "Ideal" socialism has three distinctive features:
1. Public ownership of the means of production
2. Pursuit of collective goals; and
3. Centralized decision making.

F.   **A mixed economy** combines elements of a market economy (capitalism) with elements of a command economy (socialism).

III.   PERSPECTIVES ON ECONOMY AND WORK IN THE UNITED STATES

A.   From a functionalist perspective, the economy is a vital social institution because it is the means by which goods and services are produced and distributed.

    1.   When the economy runs smoothly, other parts of society function more effectively; however, if the system becomes unbalanced, a maladjustment occurs.

    2.   The business cycle is the rise and fall of economic activity relative to long-term growth in the economy.

B.   From a conflict perspective, business cycles are the result of capitalist greed; in order to maximize profits, capitalists suppress the wages of workers.

    1.   As the prices of the products increase, the workers are not able to purchase them in the quantities that have been produced.

    2.   Consequently, surpluses occur that cause capitalists to reduce production, close factories, lay off workers, and thus contribute to the growth of the reserve army of the unemployed which then helps to reduce the wages of the remaining workers. In some situations, workers are replaced with machines or nonunionized workers.

IV.   THE SOCIAL ORGANIZATION OF WORK

A.   Sociologists who focus on microlevel analysis are interested in how the economic system and the social organization of work affect people's attitudes and behavior. Interactionists examine factors that contribute to job satisfaction or feelings of alienation.

B.   Job Satisfaction and Alienation

    1.   Job satisfaction refers to an attitude that people experience about their work, which results from (a) their job responsibilities, (b) the organizational structure in which they work, and (c) their individual needs and values.

    2.   Self-actualization occurs when people feel a sense of accomplishment and fulfillment as a result of their work. Studies have found that worker satisfaction is highest when employees have some degree of control over their work and are not too closely supervised.

    3.   Alienation occurs when workers' needs for identity and meaning are not met, and when work is done

strictly for material gain, not a sense of personal satisfaction.

C. **Occupations** are categories of jobs that involve similar activities at different work sites.
    1. The **primary labor market** is comprised of high-paying jobs with good benefits that have some degree of security and the possibility for future advancement.
    2. The **secondary labor market** is comprised of low-paying jobs with few benefits and very little job security or possibility for future advancement.

D. **Professions** are high-status, knowledge-based occupations that have five major characteristics:
    1. Abstract, specialized knowledge
    2. Autonomy
    3. Self-regulation
    4. Authority
    5. Altruism.

E. The term "manager" often is used to refer to executives, managers, and administrators who typically have responsibility for workers, physical plants, equipment, and the financial aspects of a bureaucratic organization.
    1. Scientific Management (Taylorism) was developed by industrial engineer **Frederick Winslow Taylor** to increase productivity in factories by teaching workers to perform a task in a concise series of steps; paying workers only for the number of units they produced also contributed to the success of Taylorism.
    2. Mass Production through Automation (Fordism) incorporated hierarchical authority structures and scientific management techniques into the manufacturing process. Assembly lines, machines, and robots became a means of technical control over the work process.

F. Lower Tier of the Service Sector and Marginal Jobs.
    1. Positions in the lower tier of the service sector are part of the **secondary labor market**, characterized by low wages, little job security, few chances for advancement, higher unemployment, and very limited (if any) unemployment benefits.
    2. Examples include janitors, waitresses, messengers, lower-level sales clerks, typists, file clerks, migrant laborers, and textile workers.
    3. Many jobs in this sector are **marginal jobs** which differ from the employment norms of the society in

which they are located. In the U.S., these norms are: (a) job content is legal; (b) the job is covered by government work regulations; (c) the job is relatively permanent; and (d) the job provides adequate pay with sufficient hours of work each week to make a living.

    4.   More than 11 million workers are employed in personal service industries, such as eating and drinking places, hotels, laundries, beauty shops, and household service, primarily maid service.

  G.   **Contingent work** is part-time work, temporary work, and subcontracted work that offers advantages to employers but often is detrimental to the welfare of workers.

    1.   Employers benefit by hiring workers on a part-time or temporary basis; they are able to cut costs, maximize profits, and have workers available only when they need them.

    2.   **Subcontracting** -- a form of economic organization in which a larger corporation contracts with other (usually smaller) firms to provide specialized components, product, or services -- is another form of contingent work that cuts employers costs but often at the expense of workers.

V.   UNEMPLOYMENT

  A.   There are three major types of unemployment: (1) cyclical unemployment occurs as a result of lower rates of production during recessions in the business cycle; (2) seasonal unemployment results from shifts in the demand for workers based on conditions such as weather or seasonal demands such as holidays and summer vacations; and (3) structural unemployment which arises because the skills demanded by employers do not match the skills of the unemployed or because the unemployed do not live where the jobs are located.

  B.   The **unemployment rate** is the percentage of unemployed persons in the labor force actively seeking jobs.

  C.   Unemployment compensation provides unemployed workers with short-term income while they look for other jobs.

VI.   WORKER RESISTANCE AND ACTIVISM

  A.   Labor Unions

    1.   U.S. labor unions have been credited with gaining an eight-hour work day, a five day work week, health and retirement benefits, sick leave and unemployment insurance, and workplace health and safety standards for many employees through **collective bargaining** --

negotiations between labor union leaders and employers on behalf of workers.

   2.   Although the overall number of union members in the United States has increased since the 1960s, the proportion of all employees who are union members has declined.

B.   Absenteeism, Sabotage, and Resistance
   1.   Absenteeism is one means by which workers resist working conditions they consider to be oppressive.
   2.   Other workers use sabotage to bring about informal work-stoppages, such as "throwing a monkey wrench in the gears" to halt the movement of the assembly line.
   3.   While most workers do not sabotage machinery, a significant number do resist what they perceive to be oppression from supervisors and employers.

VII.   THE GLOBAL ECONOMY IN THE TWENTY-FIRST CENTURY
A.   The U.S. economy will experience dramatic changes in the next century as workers may find themselves fighting for a larger piece of an ever shrinking economic pie which includes trade deficit and a national debt of over $5 trillion.
B.   Workers increasingly may be fragmented into two major divisions -- those who work in the innovative, primary sector and those whose jobs are located in the growing secondary, marginal sector of the labor market.
C.   Most futurists predict that multinational corporations will become even more significant in the global economy of the twenty-first century, and these corporations will become even less aligned with the values of any one nation.

ANALYZING AND UNDERSTANDING THE BOXES

After reading the chapter and studying the outline, re-read the four boxes and write down key points and possible questions for class discussion.

Sociology and Everyday Life -- "How Much Do You Know About the Economy and the World of Work?"

Key Points:

Discussion Questions:

1.

2.

3.

Sociology and Media -- "Labor Unions in the Cartoon World"

Key Points:

Discussion Questions:

1.

2.

3.

Sociology and Law -- "Labor, Business, and Government"

Key Points:

Discussion Questions:

1.

2.

3.

Sociology in Global Perspective -- "Women and Labor Activism"

Key Points:

Discussion Questions:

1.

2.

3.

PRACTICE TEST

MULTIPLE CHOICE QUESTIONS

Select the response that best answers the question or completes the statement:

1. According to Karl Marx, workers experience alienation because: (pp. 425-425)
    a.   they do not own the end product that they help create.
    b.   their jobs continue to expand.
    c.   their part in the overall production process is clearly spelled out to them by their bosses.
    d.   their tasks have become more diverse.

2. The owner of a large corporation has amassed several million dollars that she utilizes in expanding her enterprise. This wealth is referred to as: (p. 427)
    a.   services.
    b.   labor.
    c.   goods.
    d.   capital.

3. All of the following are examples of primary sector production, except: (p. 432)
    a.   oil wells.
    b.   coal mining.
    c.   textiles.
    d.   logging.

4. Industrial economies are characterized by _____ sector production. (p. 429)
    a.   primary
    b.   secondary
    c.   tertiary
    d.   quartiary

5. In sociologist Daniel Bell's vision of the future, the U.S. economy will be primarily characterized by: (p. 431)
   a. agriculture.
   b. manufacturing.
   c. service and information processing.
   d. blue-collar, working class jobs.

6. According to the text, all of the following are features of capitalism, except: (p. 432)
   a. private ownership of the means of production.
   b. pursuit of personal profit.
   c. competition.
   d. governmental intervention in the marketplace.

7. Huge corporations developed during the period of _____. (p. 433)
   a. early capitalism.
   b. beginning monopoly capitalism.
   c. advanced monopoly capitalism.
   d. post-capitalism.

8. If Sony, Philips, Time Warner, and only a few other companies control the entire music industry, this arrangement would be a(n): (p. 437)
   a. oligopoly.
   b. monopoly.
   c. conglomerate.
   d. amalgamation.

9. _____ is an economic system characterized by public ownership of the means of production, the pursuit of collective goals, and centralized decision making. (p. 440)
   a. Capitalism
   b. Communism
   c. Socialism
   d. Communitarianism

10. Democratic socialism is characterized by: (p. 442)
    a. extensive government action to provide support and services to its citizens.
    b. private ownership of the means of production, from which personal profits can be derived through market competition and without government intervention.
    c. public ownership of the means of production, the pursuit of collective goals, and centralized decision making.
    d. private ownership of some of the means of production, governmental distribution of some essential goods and services, and free elections.

11. According to the _____ perspective, when the economy runs smoothly, other parts of society function more effectively. (p. 443)
    a. conflict
    b. functionalist
    c. interactionist
    d. neo-Marxist

12. Job satisfaction refers to people's attitudes toward their work, based on: (p. 444)
    a. their job responsibilities.
    b. the organizational structure in which they work.
    c. their individual needs and values.
    d. all of the above.

13. Studies have found that job satisfaction is highest when workers: (p. 444)
    a. have some degree of control over their work.
    b. are removed from the decision-making process.
    c. know that their supervisors closely watch all aspects of their work.
    d. do not feel the pressure of playing an important part in the outcome.

14. All of the following are characteristics of professions, except: (p.446)
    a. broad-based knowledge on a wide variety of topics.
    b. authority.
    c. concern for others, not just self-interest.
    d. self-regulation.

15. Attorneys who delegate the completion of standard forms and documents to paralegals or legal secretaries are exhibiting which of the following characteristics of professions? (p. 446)
    a.   broad-based knowledge on a wide variety of topics.
    b.   authority.
    c.   concern for others, not just self-interest.
    d.   self-regulation.

16. Scientific management (Taylorism) is characterized by: (p. 448)
    a.   mass production.
    b.   automation.
    c.   time-and-motion studies.
    d.   Fordism.

17. Fast-food restaurants like McDonald's and Burger King illustrate: (p. 449)
    a.   the industrial society. (p. 449)
    b.   the piece-rate system.
    c.   Fordism.
    d.   robotics.

18. According to the text, _____ jobs differ from employment norms of the society in which they are located. (p. 450)
    a.   primary tier
    b.   marginal
    c.   peripheral
    d.   criminal

19. Studies of immigrant women workers in the Silicon Valley have concluded that: (p. 453)
    a.   capitalist economic development in the past decade has been relatively free of racism as a method of labor division and control.
    b.   these workers recently have joined unions and are now demanding higher wages.
    c.   women currently make up nearly 100 percent of the high-tech workforce in California.
    d.   gender segregation is apparent on high-tech assembly lines.

20. Capital flight refers to: (p. 454)
   a.   the investment of capital in foreign facilities.
   b.   workers fleeing their jobs for a more leisurely lifestyle in rural areas.
   c.   high-level administrators asking to be reassigned to non-supervisory positions.
   d.   seasonal unemployment that occurs when the tourist season is over in states such as Florida and California.

TRUE-FALSE QUESTIONS

T   F   1.   The economy is the social institution responsible for the production, distribution, and consumption of goods and services. (p. 426)

T   F   2.   Preindustrial economies typically are based on the extraction of raw materials and natural resources from the environment.

T   F   3.   U.S. labor unions came into existence in the twentieth century when workers became tired of toiling for the benefit of capitalists. (p. 432)

T   F   4.   A conglomerate exists when four or fewer companies supply 50 percent or more of a particular market. (p. 437)

T   F   5.   According to Karl Marx, socialism and communism are virtually identical. (p. 440)

T   F   6.   Labor unions have won substantial benefits for workers in the United States. (p. 441)

T   F   7.   According to a functionalist perspective, some problems in society are linked to peaks and troughs in the business cycle. (p. 443)

T   F   8.   Conflict theorists focus on how the economic system and the social organization of work affect people's attitudes and behavior, especially their level of job satisfaction. (p. 444)

T   F   9.   Contemporary workers can best be classified as blue collar and white collar workers. (p. 445)

T F 10. Sociologists categorize most doctors, engineers, lawyers, professors, computer scientists, and certified public accounts as "professionals." (p. 446)

T F 11. Children whose parents are professionals are more likely to become professionals themselves. (p. 447)

T F 12. Positions in the lower tier of the service sector are part of the secondary labor market, characterized by low wages, little job security, few chances for advancement, higher unemployment rates, and very limited unemployment benefits. (p. 450)

T F 13. Private household workers are among the most powerless because they cannot rely on any resources to protect them from abuse. (p. 451)

T F 14. Temporary workers are one of the slowest growing segments of the contingent work force. (p. 453)

T F 15. Unemployment statistics do not include unemployed workers who become discouraged and no longer actively seek employment. (p. 454)

SOCIOLOGY IN OUR TIMES: DIVERSITY ISSUES

1. Have labor unions historically contributed to the growth of a middle class in the United States? If labor unions decline in the future, will this have any impact on the size of the middle class in the twenty-first century?

2. According to the text, "race and gender are factors in access to the professions." (p. 447) Do you agree or disagree with this statement? Can you give examples from your personal experience that tend to confirm or deny the statement?

3. Why are large numbers of young people, people of color, recent immigrants, and white women overrepresented in marginal jobs and other positions in the lower tier of the service sector? Do employed college students sometimes see themselves as temporarily "stuck" in jobs in their tier while many other workers see themselves as permanently "stuck" in such positions?

4.   How does resistance help people in lower-tier service jobs survive at work? (p. 457) Have you ever engaged in your own form of absenteeism, sabotage, or resistance at school or work?

5.   Do you think all workplaces of the future will reflect the increasing diversity currently occurring in the U.S. population? In your future employment, will you need to communicate with individuals whose race/ethnicity, gender, and class background may be different from your own? Are you taking steps to prepare for employment in a multicultural or global workplace? If so, how?

CHAPTER TWELVE CROSSWORD PUZZLE

For those who enjoy crossword puzzles, here is a puzzle that contains words and names from Chapter Twelve. Working the puzzle will help you in reviewing the chapter. The answers appear on page 241.

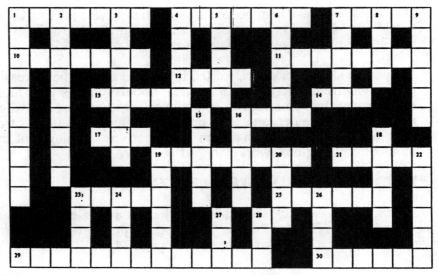

ACROSS

1. Social institution that insures the maintenance of society through the production, distribution, and consumption of [18 down] and services

4. _____ state: a state where there has been extensive government action to provide support and services to its citizens

7. A _____ of capitalism is the belief that people are free to maximize their individual gain through personal [22 down]

10. _____ labor [19 down]: consists of high-paying jobs with good benefits, some degree of security, and the possibility of advancement

11. _____ sector production: the provision of services rather than goods

12. Unemployment insurance provides unemployed workers with short-term income while they ____ for other jobs

13. _____ economy: combines elements of [30 across] and command economies

14. Conglomerates: combinations of businesses in different commercial areas, ___ of which are owned by the same holding company

16. In agrarian societies, workers have a greater variety of tasks, such as war____ and priest

17. Organizations such as the CIO changed the relationship between the working ___ and the boss

19. Generic term often used to refer to executives, administrators, etc.

21. See 5 down

23. Knights of _____: an early union

25. _____ monopoly: exists when 4 or fewer companies supply 50% or more of a particular market

29. _____ corporations are headquartered in one country and have subsidiaries in others

30. See 13 across

DOWN

1. According to the _____ norms of the U.S., a job should (among other things) provide adequate pay with sufficient hours of work to earn a living

2. This exists when several companies control an entire industry

3. Personal service and private household workers are examples of _____ jobs

4. Marx believed that, under communism, people _____ contribute according to ability and receive according to need

5. _____ union: a [21 across] of employees who join together to bargain with an employer or employers

6. Sociologists who suggested that the number of low-paying, second-tier positions in the U.S. economy has increased

7. Recession: a decline in an economy's _____ production that lasts 6 months or more

8. Esther ____-Ling Chow studied women who work in low-status occupations such as hotel housekeepers

9. He conducted time-and-motion studies, resulting in scientific management

15. First name of automaker who incorporated hierarchial authority structures and scientific management techniques

16. Corporations: large-scale organizations that have _____ powers, such as the ability to enter into contracts

18. See 1 across

20. According to one of Ben Hamper's co-workers, one of the best things about being a riveter was getting the jump on the ____ of the pack at lunchtime

22. See 7 across

23. Overall, most corporations have gained more than they have ____ as a result of government intervention

24. In the 20th century, capitalism and socialism have ____ the principal economic models in industrialized societies

26. First name of author of *The Wealth of Nations*

27. Under socialism, the people ___ the means of production

28. Initials of a major labor organization

ANSWERS TO PRACTICE TEST, CHAPTER 12

Answers to Multiple Choice Questions

1. a   According to Karl Marx, workers experience alienation because they do not own the end product that they help create. (pp. 425-425)

2. d   The owner of a large corporation has amassed several million dollars that she utilizes in expanding her enterprise. This wealth is referred to as capital. (p. 427)

3. c   All of the following are examples of primary sector production, except: textiles. (p. 432)

4. b   Industrial economies are characterized by secondary sector production. (p. 429)

5. c   In sociologist Daniel Bell's vision of the future, the U.S. economy will be primarily characterized by service and information processing. (p. 431)

6. d   According to the text, all of the following are features of capitalism, except: governmental intervention in the marketplace. (p. 432)

7. b   Huge corporations developed during the period of beginning monopoly capitalism. (p. 433)

8. a   If Sony, Philips, Time Warner and only a few other companies control the entire music industry, this arrangement would be an oligopoly. (p. 437)

9. c   Socialism is an economic system characterized by public ownership of the means of production, the pursuit of collective goals, and centralized decision making. (p. 440)

10. d   Democratic socialism is characterized by private ownership of some of the means of production, governmental distribution of some essential goods and services, and free elections. (p. 442)

11. b   According to the functionalist perspective, when the economy runs smoothly, other parts of society function more effectively. (p. 443)

12. d   Job satisfaction refers to people's attitudes toward their work, based on their job responsibilities, the organizational structure in which they work, and their individual needs and values. Consequently, "all of the above" is the best answer. (p. 444)

13. a   Studies have found that job satisfaction is highest when workers have some degree of control over their work. (p. 444)

14. a   All of the following are characteristics of professions, except: broad-based knowledge on a wide variety of topics. (p.446)

15. b   Attorneys who delegate the completion of standard forms and documents to paralegals or legal secretaries are exhibiting which of the professional characteristics? Answer: authority. (p. 446)
16. c   Scientific management (Taylorism) is characterized by time-and-motion studies. (p. 448)
17. c   Fast-food restaurants like McDonald's and Burger King illustrate Fordism. (p. 449)
18. b   According to the text, marginal jobs differ from employment norms of the society in which they are located. (p. 450)
19. d   Studies of immigrant women workers in the Silicon Valley have concluded that gender segregation is apparent on high-tech assembly lines. (p. 453)
20. a   Capital flight refers to the investment of capital in foreign facilities. (p. 454)

Answers to True-False Questions

1. True (p. 426)
2. True (p. 429)
3. False. The first national labor union came into existence in the United States in the *nineteenth century (about 1869)* when workers became tired of toiling for the benefit of capitalists. (p. 432)
4. False. A shared monopoly exists when four or fewer companies supply 50 percent or more of a particular market. (p. 437)
5. False. According to Karl Marx, communism is an economic system characterized by common ownership of all economic resources. He believed that socialism was merely a temporary stage en route to an ideal communist society. (p. 440)
6. True (p. 441)
7. True (p. 443)
8. False. Interactionists focus on how the economic system and the social organization of work affect people's attitudes and behavior, especially their level of job satisfaction. (p. 444)
9. False. Contemporary workers do not easily fit into either blue collar or white collar categories. Sociologists typically distinguish between employment in the primary and secondary labor markets instead. (p. 445)
10. True. Although many more categories of workers now refer to themselves as professionals, sociologists tend to limit the number of occupations they place in the professional middle class. (p. 446)
11. True. Children whose parents are professionals are more likely to become professionals themselves. Since higher education is one of the primary qualification for a profession, the emphasis on education

gives children whose parents are professionals a disproportionate advantage early in life. (p. 447)

12. True (p. 450)
13. True (p. 451)
14. False. Temporary workers are one of the fastest growing segments of the contingent work force. (p. 453)
15. True (p. 454)

ANSWER TO CHAPTER TWELVE CROSSWORD PUZZLE

E	C	O	N	O	M	Y	▓	W	E	L	F	A	R	E	▓	T	E	N	E	T
M	▓	L	▓	A	▓	▓	▓	O	▓	A	▓	▓	I	▓	▓	O	▓	G	▓	A
P	R	I	M	A	R	Y	▓	U	▓	B	▓	▓	T	E	R	T	I	A	R	Y
L	▓	G	▓	G	▓	▓	L	O	O	K	▓	▓	Z	▓	▓	A	▓	N	▓	L
O	▓	O	▓	M	I	X	E	D	▓	R	▓	▓	E	▓	A	L	L	▓	▓	O
Y	▓	P	▓	▓	N	▓	▓	▓	H	▓	L	O	R	D	▓	▓	▓	▓	▓	R
M	▓	O	▓	M	A	N	▓	▓	E	▓	E	▓	▓	▓	▓	G	▓	▓	▓	▓
E	▓	L	▓	▓	L	▓	M	A	N	A	G	E	R	S	▓	G	R	O	U	P
N	▓	Y	▓	▓	▓	▓	A	▓	R	▓	A	▓	E	▓	▓	▓	▓	O	▓	R
T	▓	▓	L	A	B	O	R	▓	Y	▓	L	▓	S	H	A	R	E	D	▓	O
▓	▓	▓	O	▓	E	▓	K	▓	▓	O	▓	A	T	▓	D	▓	▓	S	▓	F
▓	▓	▓	S	▓	E	▓	E	▓	▓	W	▓	F	▓	▓	A	▓	▓	▓	▓	I
M	U	L	T	I	N	A	T	I	O	N	A	L	▓	▓	M	A	R	K	E	T

CHAPTER 13
POLITICS, GOVERNMENT, AND THE MILITARY

BRIEF CHAPTER OUTLINE

Politics, Power, and Authority
> Political Science and Political Sociology
> Power and Authority

Political Systems in Global Perspective
> Monarchy
> Authoritarianism
> Totalitarianism
> Democracy

Perspectives on Power and Political Systems
> Functionalist Perspectives: The Pluralist Model
> Conflict Perspectives: Elite Models
> Critique of Pluralist and Elite Models

Politics and Government in the United States
> Political Parties
> Politics and the People

Governmental Bureaucracy
> Characteristics of the Federal Bureaucracy
> The Iron Triangle and the Military-Industrial Complex

The Military and Militarism
> Explanations for Militarism
> Race, Gender, Class, Sexual Orientation, and the Military

Political and Military Issues for the Twenty-First Century

CHAPTER SUMMARY

The relationship between politics and power is a strong one in all countries. **Politics** is the social institution through which power is acquired and exercised by some people or groups. **Power** -- the ability of persons or groups to carry out their will even when opposed by others -- is a social relationship involving both leaders and followers. Most leaders seek to legitimate their power through **authority** -- power that people accept as legitimate rather than coercive. According to Max Weber, there are three types of authority: (1) charismatic, (2) traditional, and (3) rational-legal (bureaucratic). There are four main types of contemporary political systems: monarchy, authoritarianism, totalitarianism, and democracy. In a **democracy** the people hold the ruling power either directly or through elected representatives. There are two key perspectives on how power is distributed in the United States. According to the **pluralist model**, power is widely dispersed throughout many competing interest groups. According to the

elite model, power is concentrated in a small group of elites and the masses are relatively powerless. The **power elite** is comprised of influential business leaders, key government leaders, and the military. **Political parties** are organizations whose purpose is to gain and hold legitimate control of government. The Democrats and the Republicans have dominated the U.S. political system since the mid-nineteenth century, although party loyalties have been declining in recent years. People learn political attitudes, values, and behaviors through **political socialization**. The vast governmental bureaucracy is a major source of power. The military bureaucracy is so wide-ranging that it encompasses the **military-industrial complex** -- the mutual interdependence of the military establishment and private military contractors. This complex is supported by **militarism** -- a societal focus on military ideals and an aggressive preparedness for war. Transnational trends have made is increasingly difficult for all forms of governments to control events. While these trends will become more complex in the twenty-first century, people are still likely to depend on their governments to lead and to provide solutions.

LEARNING OBJECTIVES
After reading Chapter 13, you should be able to:

1.   Explain the relationship between politics, government, and the state, and note how political sociology differs from political science.

2.   Distinguish between power and authority, and describe the three major types of authority.

3.   Compare and contrast governments characterized by monarchy, authoritarianism, totalitarianism, and democracy.

4.   State the major elements of the pluralist (functionalist) model of power and political systems.

5.   State the major elements of elite (conflict) models of power and political systems, and note how they differ from pluralist (functionalist) models.

6. Describe the purpose of political parties and analyze how well U.S. parties measure up to the ideal-type characteristics of political parties.

7. Explain the relationship between political socialization, political attitudes, and political participation.

8. Discuss the characteristics of the federal bureaucracy and explain what is meant by the "permanent government."

9. Describe the military-industrial complex and explain why it is called an iron triangle.

10. Discuss militarism and explain why support for this ideology has been so strong in the United States.

KEY TERMS (defined at page number shown and in glossary)

authoritarianism 475
authority 469
charismatic authority 470
civil rights 474
democracy 476
elite model 479
government 467
militarism 494
military-industrial complex 493
monarchy 475
pluralist model 477
political action committees 478
political party 483
political socialization 486
political sociology 469
politics 467
power 469
power elite 480
rational-legal authority 471
routinization of charisma 470
special interest groups 477
state 469
totalitarianism 475
traditional authority 470

KEY PEOPLE (identified at page number shown)

CHAPTER OUTLINE

I.   POLITICS, POWER, AND AUTHORITY

    A.   **Politics** is the social institution through which power is acquired and exercised by some people and groups.

    B.   In contemporary societies, the primary political system is the **government** -- the formal organization that has the legal and political authority to regulate the relationships among members of a society and between the society and those outside its borders.

    C.   Sociologists often refer to the government as the **state** -- the political entity that possesses a legitimate monopoly over the use of force within its territory to achieve its goals.

    D.   While political science primarily focuses on power and the distribution of power in different types of political systems, **political sociology** examines the nature and consequences of power within or between societies and focuses on the social circumstances of politics and the interrelationships between politics and social structures.

    E.   Power and Authority

        1.   **Power** is the ability of persons or groups to achieve their goals despite opposition from others.

        2.   **Authority** is power that people accept as legitimate rather than coercive.

        3.   According to Max Weber, there are three ideal types of authority:

            a.   **Charismatic authority** is power legitimized on the basis of a leader's exceptional personal qualities or the demonstration of extraordinary insight and accomplishment which inspire loyalty and obedience from followers.

            b.   **Traditional authority** is power that is legitimized on the basis of long-standing custom.

            c.   **Rational-legal authority** is power legitimized by law or written rules and regulations.

II. POLITICAL SYSTEMS IN GLOBAL PERSPECTIVE
    A.   Emergence of Political Systems
        1.   Hunting and gathering societies do not have political institutions as such because they have very little division of labor or social inequality.
        2.   Political institutions first emerge in agrarian societies as they acquire surpluses and develop greater social inequality.
        3.   Nation-states -- political organizations that have recognizable national boundaries within which their citizens possess specific legal rights and obligations -- developed first in Spain, France, and England between the twelfth and fifteenth centuries.
    B.   **Monarchy** is a political system in which power resides in one person or family and is passed from generation to generation through lines of inheritance.
    C.   **Authoritarianism** is a political system controlled by rulers who deny popular participation in government.
    D.   **Totalitarianism** is a political system in which the state seeks to regulate all aspects of people's public and private lives.
    E.   **Democracy** is a political system in which the people hold the ruling power either directly or through elected representatives.

III. PERSPECTIVES ON POWER AND POLITICAL SYSTEMS
    A.   Functionalist Perspectives: According to the **pluralist model**, power in political systems is widely dispersed throughout many competing **special interest groups** -- political coalitions comprised of individuals or groups that share a specific interest they wish to protect or advance with the help of the political system.
        1.   Key elements of this model:
            a.   Decisions are made on behalf of the people by leaders who engage in a process of bargaining, accommodation, and compromise.
            b.   Competition among leadership groups (such as leaders in business, labor, education, law, medicine, consumer groups, and government) protects people by making the abuse of power by any one group more difficult.
            c.   People can influence public policy by voting in elections, participating in existing special interest groups, or forming new ones to gain access to the political system.

> d. Power is widely dispersed in society; leadership groups that wield influence on some decisions are not the same groups which may be influential in other decisions.
>
> e. Public policy is not always based on majority preference; it is the balance between competing interest groups.

2. Over the last two decades, special interest groups have become more involved in "single-issue politics," such as abortion, gun control, gay and lesbian rights, or environmental concerns.

3. **Political action committees (PACs)** are organizations of special interest groups that fund campaigns to help elect (or defeat) candidates based on their stances on specific issues.

B. Conflict Perspectives: According to the **elite model**, power in political systems is concentrated in the hands of a small group of elites and the masses are relatively powerless.

1. Key elements of this model:

> a. Decisions are made by the elite, possessing greater wealth, education, status, and other resources than does the "masses" it governs.
>
> b. Consensus exists among the elite on the basic values and goals of society; however, consensus does not exist among most people in society on these important social concerns.
>
> c. Power is highly concentrated at the top of a pyramid-shaped social hierarchy; those at the top of the power structure come together to set policy for everyone.
>
> d. Public policy reflects the values and preferences of the elite, not the preferences of the people.

2. According to C. Wright Mills, the **power elite** is comprised of leaders at the top of business, the executive branch of the federal government, and the military (especially the "top brass" at the Pentagon). The elites have similar class backgrounds and interests.

> a. The corporate rich are the most powerful because of their unique ability to parlay the vast economic resources at their disposal into political power.

b. At the middle level of the pyramid, Mills placed the legislative branch of government, interest groups, and local opinion leaders.

c. The bottom (and widest layer) of the pyramid is occupied by the unorganized masses who are relatively powerless and vulnerable to economic and political exploitation.

3. G. William Domhoff referred to elites as the **ruling class** -- a relatively fixed group of privileged people who wield sufficient power to constrain political processes and serve underlying capitalist interests.

a. Individuals in the upper echelon are members of a business class based on the ownership and control of large corporations.

b. The intertwining of the upper class and the corporate community produces cohesion at both the economic and social levels.

c. Members of the ruling class also are linked though exclusive social clubs, expensive private schools, debutante parties, listings in the Social Register, and other upper-class indicators.

d. The corporate rich and their families influence the political process in three ways:
(1) they influence the candidate selection process by helping to finance campaigns and providing favors to political candidates;
(2) through participation in the special interest process, they are able to gain favors, tax-breaks, regulatory rulings, and other governmental supports;
(3) they gain access to the policy-making process by holding prestigious positions on governmental advisory committees, presidential commissions, and other governmental appointments.

4. Class Conflict Perspectives

a. Most contemporary elite models are based on the work of Karl Marx; however, there are divergent viewpoints about the role of the state within this perspective.

b. While instrumental Marxists argue that the state acts invariably to perpetuate the capitalist class, structural Marxists contend that the

248

state is not simply a passive instrument of the capitalist class.

IV. POLITICS AND GOVERNMENT IN THE UNITED STATES
    A.   **A political party** is an organization whose purpose is to gain and hold legitimate control of government.
        1.   Parties develop and articulate policy positions; educate voters about the issues and simplify the choices for them; recruit candidates who agree with those policies, and help those candidates win office; and, when elected, hold the candidates responsible for implementing the party's policy positions.
        2.   Since the Civil War, two political parties -- the Democratic and the Republican -- have dominated the political system in the United States and confronted two broad types of concerns.
            a.   Social issues are those relating to moral judgments or civil rights.
            b.   Economic issues involve the amount that should be spent on government programs and the extent to which these programs should encourage a redistribution of income and assets.
    B.   Politics and the People
        1.   **Political socialization** is the process by which people learn political attitudes, values, and behavior. For young children, the family is the primary agent of political socialization.
        2.   Socioeconomic status affects people's political attitudes, values, and beliefs.
        3.   Political participation occurs at four levels:
            a.   voting
            b.   attending and taking part in political meetings
            c.   actively participating in political campaigns
            d.   running for or holding political office.
        4.   At most, about 10 percent of the voting age population in this country participates at a level higher than simply voting, and only slightly more than 50 percent of the voting age population voted in the 1992 presidential election.

V. GOVERNMENTAL BUREAUCRACY
    A.   Characteristics of the Federal Bureaucracy
        1.   The size and scope of government has grown in recent decades partially because of dramatic increases in technology and in demands from the public that the

government "do something" about various problems facing society.

    2.   Much of the actual functioning of the government is carried on by the permanent government in Washington which is made up of top-tier, civil-service bureaucrats who have built a major power base.

    3.   The governmental bureaucracy has been able to perpetuate itself and expand because it has many employees with highly specialized knowledge and skills who cannot easily be replaced by those from the "outside."

B.   The Iron Triangle and the Military-Industrial Complex

    1.   The iron triangle is a three-way arrangement in which a private-interest group (usually a business corporation), a congressional committee or subcommittee, and a bureaucratic agency make the final decision on a political issue that is to be decided by that agency.

    2.   A classic example of the iron triangle is an arrangement referred to as the **military-industrial complex** -- the mutual interdependence of the military establishment and private military contractors which started during World War II and has continued to the present.

        a.   Between 1947 and 1992, the U.S. government spent an estimated $10.2 trillion for national defense (in constant 1987 dollars); several of the largest multinational corporations are among the largest defense contractors, and the defense divisions of these companies have a virtual monopoly over defense contracts.

        b.   Many members of Congress have actively supported the military-industrial complex because it provided a unique chance for them to provide economic assistance for their local, voting constituencies in the form of funding for defense-related industries, military bases, and space centers in their home state.

VI.   THE MILITARY AND MILITARISM

    A.   **Militarism** is a societal focus on military ideals and an aggressive preparedness for war; militarism is supported by core U.S. values such as patriotism, courage, reverence, loyalty, obedience, and faith in authority. Sociologists have proposed several reasons for militarization:

1. The economic interests of capitalists, college and university faculty and administrators who are recipients of research grants, workers, labor union members, and others who depend on military spending.
2. The role of the nation and its inclination toward coercion in response to perceived threats.
3. The relationship between militarism and masculinity.

B. Race, Gender, Class, Sexual Orientation, and the Military
   1. The introduction of the all-volunteer force in 1973 and the end of the draft shifted the focus of the military from training "good citizen-soldiers" to an image of the "economic person" -- one who enlists in the military in the same way that a person might take a job in the private sector.
   2. Considerable pressure has been placed on the military to recruit women; however, even as some women have moved into the ranks of officers, sexual harassment and unsolicited physical contact (including rape) have continued.
   3. Militarization uses and affects women differently based on their ethnic or racial group; by 1985, African American women made up 42 percent of all enlisted (nonofficer) women in the United States Army, a percentage four times their proportion compared with all United States women.
   4. People of color may enlist in the military, not because of their militaristic tendencies, but because of the more limited options available to them in society at large, especially if they come from lower-income family backgrounds.
   5. Discrimination against lesbians and gay men also is a pressing social issue in the U.S. military services.

VII. POLITICAL AND MILITARY ISSUES FOR THE TWENTY-FIRST CENTURY
   A. New challenges make it increasingly difficult for nations to control events such as the proliferation of arms and nuclear weapons and domestic or international terrorism.
   B. International agencies, such as the United Nations, the World Bank, and the International Monetary Fund, face many of the same problems that individual governments do -- including severe economic constraints and extreme differences of opinion among participants.
   C. While some people believe that the answer for democracy in the twenty-first century is greater participation in politics,

others believe that the current system will not change as long as governmental bureaucracies have the ability to set their own rules and regulations and monitor the everyday lives of people.

ANALYZING AND UNDERSTANDING THE BOXES

After reading the chapter and studying the outline, re-read the four boxes and write down key points and possible questions for class discussion.

Sociology and Everyday Life -- "How Much Do You Know About Politics, the Military, and Sexual Orientation?"

Key Points:

Discussion Questions:

1.

2.

3.

Sociology and Law -- "The Military's Exclusionary Policy"

Key Points:

Discussion Questions:

1.

2.

3.

Sociology and Media -- "The Politics of Outing: Mainstream and Alternative Presses"

Key Points:

Discussion Questions:

1.

2.

3.

Sociology in Global Perspective -- "An International Ban on Gays and Lesbians in the Military?"

Key Points:

Discussion Questions:

1.

2.

3.

PRACTICE TEST

MULTIPLE CHOICE QUESTIONS

Select the response that best answers the question or completes the statement:

1.   The social institution through which power is acquired and exercised by some people and groups is known as: (p. 467)
       a.   government.
       b.   politics.
       c.   the state.
       d.   the military.

2. According to the text, the state is: (p. 469)
   a. the social institution through which power is acquired and exercised by some people and groups.
   b. the formal organization that has the legal and political authority to regulate the relationships among members of a society and between the society and those outside its borders.
   c. the political entity that possesses a legitimate monopoly over the use of force within its territory to achieve its goals.
   d. the political entity that seeks to regulate all aspects of people's public and private lives.

3. _____ is the power that people accept as legitimate rather than coercive. (p. 469)
   a. Influence
   b. Clout
   c. Authority
   d. Legitimation

4. Traditional authority is based on: (p. 470)
   a. a leader's exceptional personal qualities.
   b. written rules and regulations of law.
   c. documents such as the U.S. Constitution.
   d. long-standing custom.

5. Napoleon, Julius Caesar, Martin Luther King, Jr., Cesar Chavez, and Mother Teresa are examples of: (p. 470)
   a. charismatic authority.
   b. traditional authority.
   c. rational-legal authority.
   d. nontraditional authority.

6. All of the following statements regarding racialized patriarchy are correct, except: (p. 470)
   a. Zillah R. Eisenstein coined the term "racialized patriarchy."
   b. Racialized patriarchy is closely intertwined with rational-legal authority.
   c. Racialized patriarchy is the continual interplay of race and gender.
   d. Racialized patriarchy remains a reality in both preindustrial and industrialized nations.

7. Categories of people who are excluded from the military include: (p. 473)
   a. persons with certain types of disabilities.
   b. single parents.
   c. persons who acknowledge being gay or lesbian.
   d. all of the above.

8. A hereditary right to rule or a divine right to rule is most likely to be found in a(n): (p. 475)
   a. monarchy.
   b. authoritarian regime.
   c. totalitarian regime.
   d. democracy.

9. The National Socialist (Nazi) party in Germany during World War II is an example of a(n): p. 475)
   a. monarchy.
   b. authoritarian regime.
   c. totalitarian regime.
   d. democracy.

10. All of the following statements regarding democracy are true, <u>except</u>: (p. 476)
    a. Democracy is a political system in which the people hold the ruling power either directly or through elected representatives.
    b. Several nations have attempted direct democracy at the national level.
    c. Representative democracy is not always equally accessible to all people in a nation.
    d. The framers of the Constitution established a system of representative democracy in the United States.

11. According to the pluralist model, power in political systems is: (p. 477)
    a. widely dispersed throughout many competing interest groups.
    b. concentrated in the hands of a small group of elites.
    c. comprised of leaders at the top of business, the executive branch of the federal government, and the military.
    d. controlled by members of the ruling class.

12.   According to the text, one of the reasons why lesbians and gay men may be excluded from the military is: (p. 479)
       a.   heterophobia.
       b.   homophobia.
       c.   gender bias.
       d.   militarism.

13.   The power elite model was developed by: (p. 480)
       a.   Emile Durkheim.
       b.   Thomas R. Dye and Harmon Zeigler.
       c.   Joseph Steffan.
       d.   C. Wright Mills.

14.   Political action committees: (p. 478)
       a.   generally have been abolished by reforms in campaign finance laws.
       b.   are comprised of people who volunteer their time but not money to political candidates and parties.
       c.   are organizations of special interest groups that fund campaigns to help elect candidates based on their stances on specific issues.
       d.   encourage widespread political participation by citizens at the grassroots level.

15.   The organizations responsible for developing and articulating policy positions, educating voters about issues, and recruiting candidates to run for office are known as: (p. 483)
       a.   political parties.
       b.   political action committees.
       c.   interest groups.
       d.   federal election committees.

16.   _____ is the process by which people learn political attitudes, values, and behavior. (p. 486)
       a.   Indoctrination
       b.   Military training
       c.   Resocialization
       d.   Political socialization

17.   According to the text, most of the actual functioning of the U.S. government is carried on by _____. (p. 489)
       a.   political action committees
       b.   the federal bureaucracy.
       c.   the criminal justice system.
       d.   political parties.

18. The three-way arrangement in which a private interest group, a congressional committee, and a bureaucratic agency make the final decision on a political issue that is to be decided by that agency is known as: (p. 492)
    a. political subversion.
    b. the iron law of oligarchy.
    c. the iron triangle of power.
    d. the power elite.

19. _____ is a societal focus on military ideals and an aggressive preparedness for war. (p. 494)
    a. Militarism
    b. Authoritarianism
    c. Totalitarianism
    d. Warmongerism

20. Which of the following statements is <u>correct</u> regarding the U.S. military? (p. 495)
    a. As more women have moved into the ranks of officers in the military, sexual harassment largely has been eliminated.
    b. Women's roles in the military have disturbed the existing world of male militarism.
    c. The percentage of African American women enlisted in the U.S. Army is much smaller than their proportion in the U.S. population.
    d. People of color and poor people from all racial-ethnic groups may enlist in the military because of the more limited options available to them in society at large.

TRUE-FALSE QUESTIONS

T   F   1.   Political sociology is the same as political science. (p. 469)

T   F   2.   Power is a social relationship that involves both leaders and followers. (p. 469)

T   F   3.   Charismatic authority tends to be temporary and relatively unstable. (p. 470)

T   F   4.   A nation-state is a unit of political organization that has recognizable national boundaries and whose citizens possess specific legal rights and obligations. (p. 474)

T F 5. In a representative democracy, elected representatives are expected to keep the "big picture" in mind, and they do not necessarily need to convey the concerns of those they represent in all situations. (p. 476)

T F 6. Functionalists suggest that divergent viewpoints lead to a system of political pluralism in which the government functions as an arbiter between competing interests and viewpoints. (p. 477)

T F 7. Over the past two decades, special interest groups have become more involved in multiple-issue politics because so many political issues are intertwined with other concerns. (p. 478)

T F 8. Outing is the deliberate revelation by lesbians and gays of the hidden homosexuality of prominent people. (p. 481)

T F 9. Sociologist G. William Domhoff believes that the ruling class wields sufficient power to constrain political processes in the United States. (p. 482)

T F 10. Political parties are usually composed of people with similar attitudes, interests, and socioeconomic status. (p. 483)

T F 11. Children do not typically identify with the political party of their parents until they reach college-age. (p. 486)

T F 12. The United States has one of the highest percentages of voter turnout of all Western nations. (p. 487)

T F 13. The U.S. governmental bureaucracy has been able to perpetuate itself and expand because many of its employees have highly specialized knowledge and skills and cannot be replaced easily by "outsiders." (pp. 489-490)

T F 14. The term "military-industrial complex" was first used by G. William Domhoff. (p. 493)

T   F   15.   According to interactionists, militarism may be related
                 to the social construction of masculinity in societies.
                 (p. 495)

SOCIOLOGY IN OUR TIMES: DIVERSITY ISSUES

1.   Why do you think Colonel Margarethe Cammermeyer's outstanding
      service to the military became relatively insignificant when she stated
      to the military interviewer, "I am a lesbian?" (p. 465) Do you think
      the military should be able to use sexual orientation as criteria for
      excluding people from military service? Why or why not?

2.   Can you think of other white women and people of color who might
      be considered to have charismatic authority today? Why do you think
      it is more difficult for subordinate group members than dominant
      group members in a society to be viewed as charismatic leaders?

3.   Are you concerned about your civil rights in the future? Is it
      important to protect everyone's civil rights if we hope to protect our
      own? Why or why not?

4.   How does an emphasis on militarism affect men in a society? How
      does it affect women? What is the relationship between race, gender,
      class, sexual orientation, and the military as we approach the
      twenty-first century?

CHAPTER THIRTEEN CROSSWORD PUZZLE

For those who enjoy crossword puzzles, here is a puzzle that contains
words and names from Chapter Thirteen. Working the puzzle will help you
in reviewing the chapter. The answers appear on page 263.

ACROSS

1. The power [22 down] people accept as legitimate rather than coercive
5. Political system in which power resides in [27 down] person or family and is passed from generation to generation through lines of inheritance
8. The power elite is composed of persons at the _____ of business, the executive branch of the federal government, and the military
11. _____ authority: power that is legitimized on the basis of long-standing custom
12. _____ rights: [17 down] ability of all citizens to participate as equals in the practices of democratic life
13. In most elections over the past 100 years, the voters have opted to _____ either the Democratic or the Republican candidate for President
14. The political entity that possesses a legitimate monopoly over the use of force within its territory to achieve its goals
16. Political _____: an organization whose purpose is to gain and hold legitimate control of the government
18. Bureaucratic authority: power legitimized by _____ or written rules and regulations
19. Militarism: a societal focus on military ideals and an aggressive preparedness for _____
20. _____ model: power in political systems is concentrated in the hands of a small group of elites, and the masses are relatively powerless
24. In [20 across], the _____ power is in the hands of the small group of elites
25. _____ authority: power legitimized on the basis of a leader's exceptional personal qualities

DOWN

1. Political _____ committees
2. She is one of the examples cited in the text of persons with the type of authority described in 24 across
3. _____-legal authority is also known as bureaucratic authority
4. Government: formal organization that has the legal and political authority [15 down] regulate the relationships among members of a society and between the society and _____ outside its borders
5. _____-industrial complex
6. In [3 down], authority is invested in the office and _____ the person who holds the office
7. Considering that 42% of all enlisted women in the U.S. Army are African American, it can be concluded that the U.S. military's appeal to or claim on women is far from _____ neutral
9. _____ socialization: the process by which people learn political attitudes, values, and behavior
10. Position held by the ultimate leader of a totalitarian government
15. See 4 down
17. See 12 across
21. Initials for organization of special interest groups that fund campaigns to help elect (or defeat) candidates based on their stance on specific issues
22. See 1 across
23. Last 4 letters in name of one country listed in the text as having an authoritarian government; this also is a word in the English language
26. To carry out its goals, [16 across] must recruit candidates to _____ for political office
27. See 5 across
28. Pluralist model: power in political systems _____ widely dispersed throughout many competing interest groups

ANSWERS TO PRACTICE TEST, CHAPTER 13

Answers to Multiple Choice Questions

1. b   The social institution through which power is acquired and exercised by some people and groups is known as politics. (p. 467)

2. c   According to the text, the state is the political entity that possesses a legitimate monopoly over the use of force within its territory to achieve its goals. (p. 469)

3. c   Authority is the power that people accept as legitimate rather than coercive. (p. 469)

4. d   Traditional authority is based on long-standing custom. (p. 470)

5. a   Napoleon, Julius Caesar, Martin Luther King, Jr., Cesar Chavez, and Mother Teresa are examples of charismatic authority. (p. 470)

6. b   All of the following statements regarding racialized patriarchy are correct, except: racialized patriarchy is closely intertwined with rational-legal authority. Instead, racialized patriarchy is closely intertwined with traditional authority. (p. 470)

7. d   Categories of people who are excluded from the military include: persons with certain types of disabilities, single parents, and persons who acknowledge being gay or lesbian. Thus, "all of the above" is the best answer. (p. 473)

8. a   A hereditary right to rule or a divine right to rule is most likely to be found in a monarchy. (p. 475)

9. c   The National Socialist (Nazi) party in Germany during World War II is an example of a totalitarian regime. (p. 475)

10. b   All of the following statements regarding democracy are TRUE, except: several nations have attempted direct democracy at the national level. (p. 476)

11. a   According to the pluralist model, power in political systems is widely dispersed throughout many competing interest groups. (p. 477)

12. b   According to the text, one of the reasons why lesbians and gay men may be excluded from the military is homophobia. (p. 479)

13. d   The power elite model was developed by C. Wright Mills. (p. 480)

14. c   Political action committees are organizations of special interest groups that fund campaigns to help elect candidates based on their stances on specific issues. (p. 478)

15. a The organizations responsible for developing and articulating policy positions, educating voters about issues, and recruiting candidates to run for office are known as political parties. (p. 483)

16. d Political socialization is the process by which people learn political attitudes, values, and behavior. (p. 486)

17. b According to the text, most of the actual functioning of the U.S. government is carried on by the federal bureaucracy. (p. 489)

18. c The three-way arrangement in which a private interest group, a congressional committee, and a bureaucratic agency make the final decision on a political issue that is to be decided by that agency is known as the iron triangle of power. (p. 492)

19. a Militarism is a societal focus on military ideals and an aggressive preparedness for war. (p. 494)

20. d Which of the following statements is correct regarding the U.S. military? Answer: People of color and poor people from all racial-ethnic groups may enlist in the military because of the more limited options available to them in society at large. (p. 495)

Answers to True-False Questions

1. False -- Political sociology focuses on the social circumstances of politics and explores the interrelationships between politics and social structures. Political scientists primarily focus on power and its distribution in different types of political systems. (p. 469)

2. True (p. 469)

3. True (p. 470)

4. True (p. 474)

5. False. In a representative democracy, elected representatives are expected convey the concerns and interests of those they represent. (p. 476)

6. True (p. 477)

7. False. Over the past two decades, special interest groups have become more involved in single-issue politics. (p. 478)

8. True (p. 481)

9. True (p. 482)

10. True (p. 483)

11. False. By the time children reach school age (about 5 or 6), they typically identify with the political party (if any) of their parents.

12. False. The United States has one of the lowest percentages of voter turnout of all Western nations. (p. 487)

13. True (pp. 489-490)

14. False. The term "military-industrial complex" was first used by President Dwight D. Eisenhower in his farewell address. The term also is closely associated with C. Wright Mills's power elite theory. (p. 493)
15. True (p. 495)

ANSWER TO CHAPTER THIRTEEN CROSSWORD PUZZLE

```
A U T H O R I T Y    M O N A R C H Y    T O P
C    H    A    H    I    O    A       D    O
T R A D I T I O N A L    T    C    C I V I L
I    T    I    S    I       E    C       I
O    C H O O S E    S T A T E    P A R T Y    T
N    H    N    T    A    O    O    A       I
     E    A    H    R    L A W    T       C
W A R    E L I T E    Y    P    E    O    A
E    T       W    A    R    R E A L
B    C H A R I S M A T I C       O
E    A    U    I          I    N
R O U T I N I Z A T I O N    S P E C I A L
```

CHAPTER 14
FAMILIES AND INTIMATE RELATIONSHIPS

BRIEF CHAPTER OUTLINE
Families in Global Perspective
      Family Structure and Characteristics
      Marriage Patterns
      Descent and Inheritance
      Power and Authority in Families
      Residential Patterns
Perspectives on Families
      Functionalist Perspectives
      Conflict and Feminist Perspectives
      Interactionist Perspectives
U.S. Families and Intimate Relationships
      Developing Intimate Relationships
      Cohabitation and Domestic Partnerships
      Marriage
      Housework
      Parenting
Transitions in Families
      Divorce
      Remarriage
Diversity in Families
      Diversity Among Singles
      African American Families
      Latina/o Families
      Asian American Families
      Native American Families
Family Issues in the Twenty-First Century

CHAPTER SUMMARY
Families are relationships in which people live together with commitment, form an economic unit and care for any young, and consider their identity to be significantly attached to the group. While the **family of orientation** is the family into which a person is born and in which early socialization usually takes place, the **family of procreation** is the family a person forms by having or adopting children. Sociologists investigate marriage patterns (such as **monogamy** and **polygamy**), descent and inheritance patterns (such as **patrilineal, matrilineal**, and **bilateral** descent), familial power and authority (such as **patriarchal, matriarchal**, and **egalitarian** families), residential patterns (such as **patrilocal, matrilocal**, and **neolocal residence**, and

in-group or out-group marriage patterns (i.e. **endogamy** and **exogamy**). Functionalists emphasize that families fulfill important societal functions, including sexual regulation, socialization of children, economic and psychological support, and the provision of social status. By contrast, conflict and feminist perspectives view the family as a source of social inequality and focus primarily on the problems inherent in relationships of dominance and subordination. Interactionists focus on family communication patterns and subjective meanings that members assign to everyday events. Families have changed dramatically in the United States where there have been significant increases in cohabitation, domestic partnerships, dual-earner marriages, single-parent families, and rates of divorce and remarriage. Divorce has contributed to greater diversity in family relationships, including stepfamilies or blended families and the complex binuclear family. While some never-married singles choose to remain single, others do so out of necessity. Support systems and extended family networks are important in African American, Latina/o, Asian American, and Native American families; however, factors such as age and class may reduce such family ties. Although all families have certain characteristics in common, each family is unique.

LEARNING OBJECTIVES
After reading Chapter 14, you should be able to:
1.	Explain why it has become increasingly difficult to develop a concise definition of family.

2.	Describe kinship ties and distinguish between families of orientation and families of procreation.

3.	Compare and contrast extended and nuclear families.

4.	Describe the different forms of marriage found across cultures.

5.	Discuss the system of descent and inheritance, and explain why such systems are important in societies.

6.   Distinguish between patriarchal, matriarchal, and egalitarian families.

7.   Explain the differences in residential patterns and note why most people practice endogamy.

8.   Describe functionalist, conflict and feminist, and interactionist perspectives on families.

9.   Describe how U.S. families have changed over the past two decades.

10.   Describe cohabitation and domestic partnerships and note key social and legal issues associated with each.

11.   Describe the major problems faced in dual-earner marriages, and note why the double shift most often is a problem for women.

12.   Discuss the major issues associated with adoption, teenage pregnancies, single-parent households, and two-parent households.

13.   Explain the major causes and consequences of divorce and remarriage in the United States.

14.   Describe the diversity found in contemporary U.S. families.

KEY TERMS (defined at page number shown and in glossary)

bilateral descent 511
cohabitation 520
domestic partnerships 521
dual-earner marriages 522
egalitarian families 511
endogamy 512
exogamy 512
extended family 508
families 505
family of orientation 507
family of procreation 507
homogamy 522
kinship 506
marriage 508

matriarchal family 511
matrilineal descent 510
matrilocal residence 512
monogamy 508
neolocal residence 512
nuclear family 508
patriarchal family 511
patrilineal descent 510
patrilocal residence 512
polyandry 510
polygamy 510
polygyny 510
second shift 522
sociology of family 512

KEY PEOPLE (identified at page number shown)

Judy Root Aulette 508
Peter Berger and
    Hansfried Kellner 515
Francesca Cancian 519
Andrew Cherlin 531
Emile Durkheim 512
Arlie Hochschild 522
Alfred C. Kinsey 520
Charlene Maill 523
Sara McLanahan and
    Karen Booth 525

Talcott Parsons 512
Brian Robinson 525
Alice Rossi 526
Lillian Rubin 534
Lenore Walker 517
Kath Weston 508
Jane Riblett Wilkie 515
Norma Williams 533

CHAPTER OUTLINE

I.   FAMILIES IN GLOBAL PERSPECTIVE
    A.   **Families** are relationships in which people live together with commitment, form an economic unit and care for any young, and consider their identity to be significantly attached to the group.
    B.   Family Structure and Characteristics
        1.   In preindustrial societies, the primary social organization is through **kinship** -- a social network of people based on common ancestry, marriage, or adoption.
        2.   In industrialized societies, other social institutions fulfill some functions previously taken care of by kinship ties; families are responsible primarily for regulating sexual activity, socializing children, and

providing affection and companionship for family members.

3. Many of us will be members of two types of families: a **family of orientation** -- the family into which we are born or adopted and in which early socialization usually takes place, and a **family of procreation** -- the family we form by having or adopting children.

4. Extended and nuclear families

   a. An **extended family** is a family unit composed of relatives (such as grandparents, uncles, and aunts) in addition to parents and children who live in the same household.

   b. A **nuclear family** is a family composed of one or two parents and their dependent children, all of whom live apart from other relatives.

C. Marriage Patterns

1. **Marriage** is a legally recognized and/or socially approved arrangement between two or more individuals that carries certain rights and obligations and usually involves sexual activity.

2. In the United States, **monogamy** -- a marriage between two partners, usually a woman and a man -- is the only form of marriage sanctioned by law.

3. **Polygamy** is the concurrent marriage of a person of one sex with two or more members of the opposite sex.

   a. The most prevalent form of this marriage pattern is **polygyny** -- the concurrent marriage of one man with two or more women.

   b. **Polyandry**-- the marriage of one woman with two or more men -- is very rare.

D. Descent and Inheritance

1. In preindustrial societies, the most common pattern of unilineal descent is **patrilineal descent** -- tracing descent through the father's side of the family -- whereby a legitimate son inherits his father's property and sometimes his position upon the father's death.

2. **Matrilineal descent** traces descent through the mother's side of the family; however, inheritance of property and position usually is traced from the maternal uncle (mother's brother) to his nephew (mother's son).

3. In industrial societies such as the United States, kinship usually is traced through both parents;

268

bilateral descent is a system of tracing descent through both the mother's and father's sides of the family.

E. Power and Authority in Families:
1. **Patriarchal family**: a family structure in which authority is held by the eldest male (usually the father), who acts as head of household and holds power over the women and children.
2. **Matriarchal family**: a family structure in which authority is held by the eldest female (usually the mother), who acts as head of household.
3. **Egalitarian family**: a family structure in which both partners share power and authority equally.

F. Residential Patterns:
1. **Patrilocal residence**: the custom of a married couple living in the same household (or community) with the husband's family.
2. **Matrilocal residence**: the custom of a married couple living in the same household (or community) with the wife's parents.
3. In industrialized nations, most couples hope to live in a **neolocal residence**: the custom of a married couple living in their own residence apart from both the husband's and the wife's parents.

G. Endogamy and Exogamy.
1. **Endogamy** refers to cultural norms prescribing that people marry within their own social group or category.
2. **Exogamy** refers to cultural norms prescribing that people marry outside their own social group or category.

II. PERSPECTIVES ON FAMILIES
A. The **sociology of family** is the subdiscipline of sociology that attempts to describe and explain patterns of family life and variations in family structure.
B. Functionalist Perspective
1. The family is important in maintaining the stability of society and the well-being of individuals.
2. According to **Emile Durkheim**, both marriage and society involve a mental and moral fusion of individuals; division of labor contributes to greater efficiency in all areas of life.
3. **Talcott Parsons** further defined the division of labor in families: the husband/father fulfills the instrumental role (meeting the family's economic needs, making

269

important decisions, and providing leadership) while the wife/mother fulfills the expressive role (doing housework, caring for children, and meeting the emotional needs of family members).

4. Four key functions of families in advanced industrial societies are:
   a. sexual regulation
   b. socialization
   c. economic and psychological support for members
   d. provision of social status and reputation.

C. Conflict and Feminist Perspectives
   1. Families are a primary source of inequality.
   2. According to some conflict theorists, families in capitalist economies are similar to workers in a factory: women are dominated at home by men the same way workers are dominated by capitalists in factories; reproduction of children and care for family members at home reinforce the subordination of women through unpaid (and devalued) labor.
   3. Some feminist perspectives focus on patriarchy rather than class because men's domination over women existed long before private ownership of property; contemporary subordination is rooted in men's control over women's labor power.

D. Interactionist Perspectives
   1. Interactionists examine the roles of husbands, wives, and children as they act out their own parts and react to the actions of others.
   2. According to Peter Berger and Hansfried Kellner, interaction between marital partners contributes to a shared reality: newlyweds bring separate identities to a marriage but gradually construct a shared reality as a couple.
   3. According to Jessie Bernard, women and men experience marriage differently: there is "his" marriage and "her" marriage.

III. U.S. FAMILIES AND INTIMATE RELATIONSHIPS
A. Developing Intimate Relationships
   1. Although the ideal culture emphasizes romantic love, men and women may not share the same perceptions about love: women tend to express their feelings verbally, while men tend to express their love through nonverbal actions.

2. Scholars suggest that there are six dominant sexual standards in the United States:
   a. Heterosexual: attraction is limited to members of the opposite sex;
   b. Romantic: sex and love should go together;
   c. Marital: marriage should include sex;
   d. Two-person: sex must involve two (but no more than two) people;
   e. Coital: sexual intercourse should occur between a man and a woman, with coitus the ultimate sexual act; and
   f. Orgasmic: people should experience orgasm as the climax of sexual interactions; if not, something is wrong.

B. Cohabitation and Domestic Partnerships
   1. **Cohabitation** refers to a couple who live together without being legally married.
   2. Characteristics of persons most likely to cohabit are as follows: under age 45, have been married before, or are older individuals who do not want to lose financial benefits (such as retirement benefits) that are contingent upon not marrying.
   3. Many lesbian and gay couples cohabit because they cannot enter into a legally recognized marriage; some have sought recognition of **domestic partnerships** -- household partnerships in which an unmarried couple lives together in a committed sexually intimate relationship and is granted the same benefits as those accorded to married heterosexual couples.

C. Marriage
   1. Couples marry for reasons such as being "in love," desiring companionship and sex, wanting to have children, feeling social pressure, attempting to escape from their parents' home, or believing they will have greater resources if they get married.
   2. Most people engage in **homogamy** -- the pattern of individuals marrying those who have similar characteristics, such as race/ethnicity, religious background, age, education, or social class.
   3. Communication and support are crucial to the success of marriages; problems that cause the most concern are partners who are emotionally distant, poor communication, and lack of companionship.

271

D.   Housework

    1.   Over 50% of all U.S. marriages are **dual-earner marriages** -- marriages in which both spouses are in the labor force. Over half of all employed women hold full-time, year-round jobs.

    2.   Many married women also have a **second shift** -- the domestic work that employed women perform at home after they complete their workday on the job.

    3.   Women employed full-time who are single parents probably have the greatest burden of all; they have complete responsibility for the children and household, often with little or no help from ex-husbands or relatives.

E.   Parenting

    1.   Couples deciding not to have children may consider themselves "child free," while those who do not produce children through no choice of their own may consider themselves "childless."

    2.   Adoption is a legal process through which rights and duties of parenting are transferred from a child's biological and/or legal parents to new legal parents.

    3.   About 6.4 million women become pregnant each year in the United States; of that number about 44% of pregnancies are intended while 56% are unintended.

    4.   Teen pregnancies have decreased over the past three decades; however, they may be viewed as a crisis because an increase has occurred in the number of births among unmarried teenagers.

    5.   Recently, single- or one-parent households have increased significantly due to divorce and to births outside marriage.

    6.   Parenthood in the United States is idealized, especially for women.

IV.   TRANSITIONS IN FAMILIES

    A.   Divorce

        1.   Divorce is the legal process of dissolving a marriage that allows former spouses to remarry if they so choose. Most divorces are based on "irreconcilable differences" (there has been a breakdown of the marital relationship for which neither partner is specifically blamed).

        2.   Approximately 2.4 million marriages occur each year in the United States, and about 1.2 million divorces are granted. However, the couples who are divorced

in any given year are very unlikely to come from the group who married that year.

3. Causes of Divorce
   a. At the macrolevel, societal factors contributing to higher rates of divorce include changes in social institutions such as religion and law.
   b. At the microlevel, characteristics that appear to contribute to divorce are:
      (1) Marriage at an early age;
      (2) A short acquaintanceship before marriage;
      (3) Disapproval of the marriage by relatives and friends;
      (4) Limited economic resources;
      (5) Having a high-school education or less;
      (6) Parents who are divorced or have unhappy marriages;
      (7) The presence of children at the beginning of the marriage.

4. Consequences of Divorce
   a. By age 16, about one in every three white and two in every three African American children will experience divorce within their families. Some children experience more than one divorce during their childhood because one or both of their parents may remarry and subsequently divorce again.
   b. Divorce changes relationships for other relatives, especially grandparents.

B. Remarriage
   1. Most people who divorce get remarried: more than 40% of all marriages take place between previously married brides and/or grooms, and about half of all persons who divorce before age 35 will remarry within three years.
   2. Most divorced people remarry others who have been divorced. At all ages, a greater proportion of men than women remarry and often relatively soon after divorce. Among women, those who divorce at younger ages are more likely to remarry than are those who are older; women with a college degree and without children are less likely to remarry.

V. DIVERSITY IN FAMILIES
    A. Diversity Among Singles
        1. Some never-married singles choose to remain single because of opportunities for a career (especially for women), the availability of sexual partners without marriage, a belief that the single lifestyle is full of excitement, and a desire to be self-sufficient and have the freedom to change and experiment.
        2. Other never-married singles remain single out of necessity; they cannot afford to marry and set up their own household.
        3. Among persons age 18 and over in 1992, over 35% of African American women had never married, as compared with almost 24% of Latinas and 17% of whites. Among African American men, about 40% had never married as compared with about 32% of Latinos and 24% of whites.
    B. African American Families
        1. A higher proportion of African Americans than whites lives in extended family households which may provide emotional and financial support not otherwise available.
        2. Among middle and upper-middle class African American families, nuclear families are more prevalent than extended family ties.
    C. Latina/o Families
        1. Family support systems found in many Latina/o families -- "la familia" -- cover a wide array of relatives including parents, aunts, uncles, cousins, brothers and sisters, and their children.
        2. Some sociologists question the extent that familialism exists across social classes. Norma Williams found that extended family networks are disappearing, especially among advantaged urban Latinos/as.
    D. Asian American Families
        1. While many Asian Americans live in nuclear families, others (especially those residing in Chinatowns) have extended family networks. Some are referred to as semi-extended families because other relatives live in close proximity but not necessarily in the same household.
        2. Extended family networks of some Vietnamese Americans are limited because family members died in the war in Vietnam, and others did not migrate to the United States.

E.	Native American Families
    1.	Family ties remain strong among many Native Americans, who have always been known for their strong family ties, for people within one family group caring for each other.
    2.	Extended family patterns are common among lower-income Native Americans living on reservations; most others live in nuclear families.

VI.	FAMILY ISSUES IN THE TWENTY-FIRST CENTURY
    A.	Some people believe that the family as we know it is doomed; others believe that a return to traditional family values will save this social institution and create greater stability in society.
    B.	Sociologist **Lillian Rubin** suggests that clinging to a traditional image of families is hypocritical in light of our society's failure to support them. Some laws have the effect of hurting children whose families do not meet the traditional model. For example, cutting down on government programs which provide food and medical care for pregnant women and infants will result in seriously ill children rather than model families.
    C.	People's perceptions about what constitutes a family will continue to change in the next century: the family may become those persons on whom one can depend for emotional support, who are available in crises and emergencies, or who provide continuing affections, concern, and companionship.

ANALYZING AND UNDERSTANDING THE BOXES
After reading the chapter and studying the outline, re-read the four boxes and write down key points and possible questions for class discussion.

Sociology and Everyday Life -- "How Much Do You Know About Contemporary Trends in U.S. Family Life?"

Key Points:

Discussion Questions:

1.

2.

3.

Sociology and Media -- "The Simpsons: An All-American Family"

Key Points:

Discussion Questions:

1.

2.

3.

Sociology in Global Perspective -- "Family Life in Japan"

Key Points:

Discussion Questions:

1.

2.

3.

Sociology and Law -- "The Best Interests of the Children in Custody Cases"

Key Points:

Discussion Questions:

1.

2.

3.

PRACTICE TEST

MULTIPLE CHOICE QUESTIONS

Select the response that best answers the question or completes the statement:

1.   According to the text, traditional definitions of the family: (p. 504)
     a.   are still highly applicable to today's families.
     b.   include all persons in a relationship who wish to consider themselves a family.
     c.   need to be expanded to provide a more encompassing perspective on what constitutes a family.
     d.   are indistinguishable from contemporary definitions of the family.

2.   A social network of people based on common ancestry, marriage, or adoption is known as: (p. 506)
     a.   kinship.
     b.   a family.
     c.   a clan.
     d.   subculture.

3.   The family of procreation is defined as "the family _____." (p. 507)
     a.   into which a person is born
     b.   in which people receive their early socialization
     c.   that is composed of relatives in addition to parents and children who live in the same household
     d.   a person forms by having or adopting children

4.   Families that include grandparents, uncles, aunts, or other relatives who live in close proximity to the parents and children are known as a(n): (p. 508)
     a.   clan.
     b.   extended family.
     c.   nuclear family.
     d.   family of procreation.

5.   _____ is the concurrent marriage of one man with two or more women, while _____ is the concurrent marriage of one woman with two or more men. (p. 510)
     a.   Monogamy - polygamy
     b.   Patriarchy - matriarchy
     c.   Polygyny - polyandry
     d.   Polyandry - polygyny

6. The most prevalent pattern of power and authority in families is: (p. 511)
   a. matriarchy.
   b. monarchy.
   c. oligarchy.
   d. patriarchy.

7. All of the following statements are correct regarding power and authority in families, except: (p. 511)
   a. Recently, there has been a trend toward more egalitarian family relationships in a number of countries.
   b. Some degree of economic independency makes it possible for women to delay marriage.
   c. Power and authority are not important issues among gay and lesbian couples because their relationships already are more egalitarian.
   d. Scholars have found no historical evidence to indicate that true matriarchies ever existed.

8. The custom of a married couple living in their own residence apart from both the husband's and the wife's parents is known as: (p. 512)
   a. isolated residence.
   b. neolocal residence.
   c. neutral-local residence.
   d. exogamous residence.

9. When a person marries someone who comes from the same social class, racial-ethnic group, and religious affiliation, sociologists refer to this marital pattern as: (p. 512)
   a. endogamy.
   b. exogamy.
   c. inbreeding.
   d. intraclass reproduction.

10. In the United States, _____ was a key figure in developing a functionalist model of the family. (p. 512)
    a. C. Wright Mills
    b. Talcott Parsons
    c. Charles H. Cooley
    d. Peter Berger

11.   According to functionalists, all of the following are key functions of families, except: (p. 513)
        a.   provision of social status.
        b.   economic and psychological support.
        c.   maintenance of workers so that they can function effectively in the workplace.
        d.   sexual regulation and socialization of children.

12.   According to _____ theorists, interaction between marital partners contributes to a shared reality. (p. 515)
        a.   feminist
        b.   conflict
        c.   functionalist
        d.   interactionist

13.   Only one out of _____ families is composed of a married couple with one or more children under age 18. (p. 517)
        a.   two
        b.   four
        c.   twenty
        d.   fifty

14.   At what point in history did people in the United States come to view home and work as separate spheres? (p. 518)
        a.   during colonial times
        b.   during the Industrial Revolution
        c.   at the end of World War II
        d.   in the 1980s when more middle-class, white women entered the paid workforce

15.   Which of the following statements regarding cohabitation is true? (p. 520)
        a.   Attitudes about cohabitation have not changed very much in the past two decades.
        b.   The Bureau of the Census recently developed a more inclusive definition of cohabitation.
        c.   People most likely to cohabit are those who are under age 30 and who have not been married before.
        d.   In the United States, many lesbian and gay couples cohabit because they cannot enter into a legally recognized marital relationship.

16. Sociologist _____ coined the term "second shift" to refer to the domestic work that employed women perform at home after they complete their workday on the job. (p. 522)
    a. Arlie Hochschild
    b. Talcott Parsons
    c. Kath Weston
    d. Francesca Cancian

17. Recent studies of teen pregnancy have concluded that: (p. 524)
    a. teenage pregnancies are increasing rapidly in the United States.
    b. teenage mothers often are as skilled at parenting as older mothers.
    c. there has been an increase in teenage pregnancies among unmarried teenagers.
    d. the media no longer focuses on "the problem" of teenage pregnancies.

18. According to sociologist Alice Ross, men secure their status as adults by _____; women secure their status as adults by _____. (p. 526)
    a. fathering -- mothering
    b. their employment -- their employment
    c. their sports abilities -- their housekeeping skills
    d. their employment -- maternity

19. All of the following are cited in the text as primary social characteristics of those most likely to get divorced, except: (p. 527)
    a. marriage at a later age and being set in one's ways.
    b. a short acquaintanceship before marriage.
    c. disapproval of marriage by relatives and friends.
    d. parents who are divorced or have unhappy marriages.

20. In discussing diversity in families, the text points out that: (p. 531-534)
    a. a higher percentage of whites live in extended family households than do African Americans.
    b. there is no such thing as "the" African American, Latina/o, Asian American, or Native American family.
    c. working-class African American women often are encouraged by their families to chose marriage over education.
    d. extended family patterns are no longer common among lower-income Native Americans living on reservations.

TRUE-FALSE QUESTIONS

T   F   1.   Alcohol abuse is not a major problem in U.S. families. (p. 503)

T   F   2.   In industrialized societies, other social institutions fulfill some of the functions previously taken care of by the kinship network. (p. 506)

T   F   3.   By definition, "marriage" must be a legally recognized and/or socially approved arrangement. (p. 508)

T   F   4.   Polyandry brings prestige to men through their wives' work and the children they produce. (p. 510)

T   F   5.   In industrial societies, kinship is usually traced through patrilineal descent. (pp. 510-511)

T   F   6.   According to some feminist scholars, hostility and violence against women and children may be attributed to patriarchal attitudes, economic hardship, and rigid gender roles in society. (p. 511)

T   F   7.   Interactionists focus on families as a primary source of social inequality. (p. 512)

T   F   8.   Functionalists suggest that the erosion of family values may occur when religion becomes less important in everyday life. (p. 513)

T   F   9.   According to recent studies, most U.S. men are happy to relinquish their status as family breadwinner so that they are relieved of some of the pressure to make money and be a success. (p. 515)

T   F   10.   A domestic partnership is a household partnership in which an unmarried couple lives together in a committed, sexually intimate relationship that is granted the same rights and benefits as those accorded to married heterosexual couples. (p. 521)

T   F   11.   Over 50 percent of all marriages in the United States are dual-earner marriages. (p. 522)

T	F	12.	Some people feel that there is a social stigma associated with childlessness. (p. 523)

T	F	13.	Most single fathers who do not have custody of their children still play an important role in the lives of their children. (p. 525)

T	F	14.	Societal factors that contribute to higher rates of divorce include changes in social institutions, such as religion and family. (p. 527)

T	F	15.	Unlike U.S. women, employed women in Japan do not have a second shift of housework. (p. 528)

SOCIOLOGY IN OUR TIMES: DIVERSITY ISSUES

1. Which of the family patterns described in this chapter are applicable to your own family? What do you consider to be the strengths and weaknesses of the family structure in which you grew up? Did changes in that structure occur over time? If so, how did those changes affect you?

2. Why do you think "The Simpsons" has been such a popular animated television series? (p. 514) What subtle messages do we receive regarding race/ethnicity, gender, and class from television shows and other forms of popular culture?

3. According to the text, "across race and class, numerous studies have confirmed that domestic work remains primarily women's work." (p. 522) If you apply your sociological imagination to this statement, how are "family" problems related to beliefs and values embedded in the larger society?

4. Throughout the chapter, the unique problems of lesbian and gay couples are discussed. Do you think these problems will be resolved in the future? Or, do you think these problems may become worse? Do you think our society will become more or less tolerant of diverse family patterns in the future?

CHAPTER FOURTEEN CROSSWORD PUZZLE

For those who enjoy crossword puzzles, here is a puzzle that contains words and names from Chapter Fourteen. Working the puzzle will help you in reviewing the chapter. The answers appear on page 286.

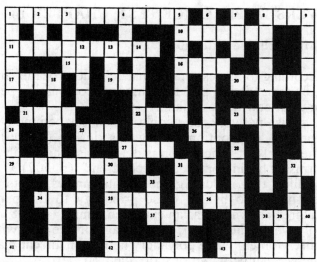

ACROSS

1. The _____ perspective emphasizes the importance of the [1 down] in maintaining stability in society
8. _____-earner marriages: both spouses are in the labor force
10. Cultural norms prescribing that people marry outside their [39 down] social groups or category
11. _____ [42 across]: a system of tracing [42 across] through the mother's side of the family
15. Serial monogamy: a succession of marriages ___ that a person has several spouses over a lifetime, but never more than one at a time
16. In matrilocal residence, the married [20 across] [17 across] in the same household as the wife's parents, or at least _____ that household
17. See 16 across
19. Monogamy: a marriage between _____ partners
20. See 16 across
21. She suggested couples should have a two-stage marriage
22. See 2 down
23. See 4 down
25. See 13 down
26. [31 down] determined that many _____ [33 down] and lesbians have "families we choose"
27. Second shift: domestic _____ that employed women perform at home after leaving their job
29. Durkheim saw marriage as a mental and moral fusion of physically _____ individuals
34. Matri_____al or patri_____al families
35. In Parsons's opinion, the _____ fulfills the expressive role
36. _____local residence: a married couple lives apart from both sets of parents
37. See 7 down
38. Bilateral descent traces kinship through _____ the mother's and the father's side of the family
41. Families form an economic unit and care for any ___
42. See 11 across
43. With Berger, noted that marriage contributes to a shared reality between the parties

DOWN

1. See 1 across
2. Scholars have ___ been able to establish that [22 across] matriarchies ever existed
3. Bilateral descent: _____ pattern is used in the U.S. for determining kinship and inheritance rights
4. Homogamy: the pattern of marrying individuals with similar characteristics such as [23 across], _____, education, or social class
5. Although _____ pregnancies are a popular topic in the media, they actually have decreased over the last 30 years
6. A couple who live together without being legally married
7. Jessie Bernard made [37 across] of the fact that a woman and a _____ experience marriage differently: his marriage and [40 down] marriage
8. Domestic violence is one _____ in some marriages
9. Talcott Parsons asserted that the husband/father fulfills the instrumental role, including providing _____
12. The U.S. emphasizes romantic _____
13. Fictive [25 across] are _____ really relatives
14. Legal process through which rights and duties of parenting are transferred to new legal parents
18. _____ family: a family structure in which both partners share power and authority equally
24. Opposite of 10 across
25. Social network of people based on common ancestry, marriage, or adoption
28. The legal process of dissolving a marriage
30. Recently, a trend _____ more [18 down] relationships has been evident
31. See 26 across
32. Extended family: a type of family _____
33. See 26 across
39. See 10 across
40 See 7 down

283

ANSWERS TO PRACTICE TEST, CHAPTER 14

Answers to Multiple Choice Questions

1.   c   According to the text, traditional definitions of the family need to be expanded to provide a more encompassing perspective on what constitutes a family. (p. 504)

2.   a   A social network of people based on common ancestry, marriage, or adoption is known as kinship. (p. 506)

3.   d   The family of procreation is defined as "the family a person forms by having or adopting children." (p. 507)

4.   b   Families that include grandparents, uncles, aunts, or other relatives who live in close proximity to the parents and children are known as an extended family. (p. 508)

5.   c   Polygyny is the concurrent marriage of one man with two or more women, while polyandry is the concurrent marriage of one woman with two or more men. (p. 510)

6.   d   The most prevalent pattern of power and authority in families is patriarchy. (p. 511)

7.   c   All of the following statements are correct regarding power and authority in families, except: power and authority are not important issues among gay and lesbian couples because their relationships already are more egalitarian. (p. 511)

8.   b   The custom of a married couple living in their own residence apart from both the husband's and the wife's parents is known as neolocal residence. (p. 512)

9.   a   When a person marries someone who comes from the same social class, racial-ethnic group, and religious affiliation, sociologists refer to this marital pattern as endogamy. (p. 512)

10.   b   In the United States, Talcott Parsons was a key figure in developing a functionalist model of the family. (p. 512)

11.   c   According to functionalists, all of the following are key functions of families, except: maintenance of workers so that they can function effectively in the workplace. (p. 513)

12.   d   According to interactionist theorists, interaction between marital partners contributes to a shared reality. (p. 515)

13.   b   Only one out of four families is composed of a married couple with one or more children under age 18. (p. 517)

14.   b   At what point in history did people in the United States come to view home and work as separate spheres? Answer: during the Industrial Revolution (p. 518)

15.   d   Which of the following statements regarding cohabitation is true? (p. 520) Answer: In the United States, many lesbian and gay couples cohabit because they cannot enter into a legally recognized marital relationship.

16. a Sociologist Arlie Hochschild coined the term "second shift" to refer to the domestic work that employed women perform at home after they complete their workday on the job. (p. 522)

17. c Recent studies of teen pregnancy have concluded that there has been an increase in teenage pregnancies among unmarried teenagers. (p. 524)

18. d According to sociologist Alice Ross, men secure their status as adults by their employment; women secure their status as adults by maternity. (p. 526)

19. a All of the following are cited in the text as primary social characteristics of those most likely to get divorced, <u>except</u>: marriage at a later age and being set in one's ways. (p. 527)

20. b In discussing diversity in families, the text points out that there is no such thing as "the" African American, Latina/o, Asian American, or Native American family. (p. 531-534)

Answer to True-False Questions

1. False -- About 7 million children live in families with at least one parent who abuses alcohol. (p. 503)
2. True (p. 506)
3. True (p. 508)
4. False -- Polygyny -- the concurrent marriage of one man with two or more women -- brings prestige to men through their wives' work and the children they produce. (p. 510)
5. False -- In industrial societies, kinship is usually traced through bilateral descent -- a system of tracing descent through both the mother's and father's sides of the family. (pp. 510-511)
6. True (p. 511)
7. False -- <u>Conflict</u> <u>theorists</u> focus on families as a primary source of social inequality. (p. 512)
8. True (p. 513)
9. False -- According to a recent study by sociologist Jane Riblett Wilkie, most U.S. men are reluctant to relinquish their status as family breadwinner. (p. 515)
10. True (p. 521)
11. True (p. 522)
12. True (p. 523)
13. False -- Most single fathers who do not have custody of their children may play a relatively limited role in the lives of their children. Sometimes this limited role is by choice, but more often, it is caused by workplace demands, location of the ex-wife's residence, and limitations placed on visitation by custody arrangements. (p. 525)
14. True (p. 527)

15. False -- Many employed Japanese women also have a second shift of housework averaging three hours a day as compared with husbands' eight minutes a day. (p. 528)

ANSWER TO CHAPTER FOURTEEN CROSSWORD PUZZLE

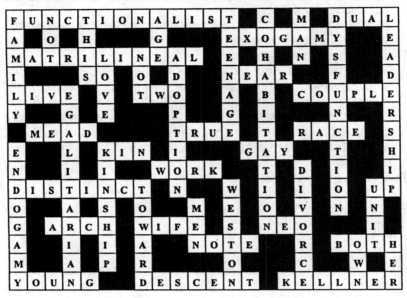

CHAPTER 15
EDUCATION AND RELIGION

BRIEF CHAPTER OUTLINE

CHAPTER SUMMARY

Education and religion are important social institutions in contemporary societies; however, there is a lack of consensus in the United States regarding the appropriate relationship between public education and religion. **Education** is the social institution responsible for the systematic transmission of knowledge, skills, and cultural values within a formally organized structure. In addition to teaching the basics, U.S. schools today also teach a myriad of topics ranging from computer skills to AIDS prevention. Functionalists have suggested that education performs a number of essential functions for society; however, conflict theorists emphasize that education perpetuates class, racial-ethnic, and gender inequalities. Interactionists point out that education may become a self-fulfilling prophecy for students who come to perform up -- or down -- to the expectations held for them by teachers. **Religion** is a system of beliefs, symbols, and rituals, based on some sacred or supernatural realm that guides human behavior, gives meaning to life, and unites believers into a community. Although religion seeks to answer important questions such as why we exist, why people suffer and die, and what happens when we die, increases in scientific knowledge have contributed to **secularization** -- the process by which religious beliefs, practices, and institutions lose their significance in sectors of society and culture. According to functionalists, religion provides meaning and purpose to life, promotes social cohesion and a sense of belonging, and provides social control and support for the government. From a conflict perspective, religion can have negative consequences in that the capitalist class uses religion as a tool of domination to mislead workers about their true interests. However, Max Weber believed that religion could be a catalyst for social change. Interactionists examine the meanings that people give to religion and the meanings they attach to religious symbols in their everyday life. Contemporary religious organizations may be categorized as ecclesia, churches, **denominations**, **sects**, and **cults**. Maintaining an appropriate balance between the social institutions of education and religion will be an important challenge for the United States in the twenty-first century.

LEARNING OBJECTIVES
After reading Chapter 15, you should be able to:
1. Trace the history of education from early societies to the present.

2. Describe the functionalist perspective on education and note the societal importance of manifest and latent functions fulfilled by this social institution.

3. Describe conflict perspectives on education and note how they differ from a functionalist perspective.

4. Discuss interactionist perspectives on education, and describe the significance of the self-fulfilling prophecy and labeling on educational achievement.

5. Trace the history of religion from early societies to the present.

6. Describe the functionalist perspective on religion and discuss its major functions in societies.

7. Compare and contrast civil religion with other forms of religion.

8. Describe conflict perspectives on religion and distinguish between the approaches of Karl Marx and Max Weber.

9. Describe interactionist perspectives on religion and explain how religion may be viewed differently by women and men.

10. Distinguish between the different types of religious organizations and note why a religious group may move from one category to another over time.

11. Describe recent trends in religion in the United States and explain their potential impact on public education in the future.

KEY TERMS (defined at page number shown and in glossary)

animism 557
church 565
civil religion 562
credentialism 549
cult 567
cultural capital 548
denomination 566
ecclesia 565
education 543
hidden curriculum 549
latent functions 547
manifest functions 546
meritocracy 549

monotheism 557
polytheism 557
profane 556
religion 556
rituals 556
sacred 556
sect 566
secularization 559
self-fulfilling prophecy 554
simple supernaturalism 557
theism 557
tracking 548
transcendent idealism 557

KEY PEOPLE (identified at page number shown)

Robert Bellah 562
Pierre Bourdieu 548
Emile Durkheim 545, 556, 560
Clifford Geertz 556
Richard J. Herrnstein and
  Charles Murray 554

Karl Marx 563
Meredith B. McGuire 559, 564
Robert Merton 554
Wade Clark Roof 560
Max Weber 563

CHAPTER OUTLINE

I.   AN OVERVIEW OF EDUCATION AND RELIGION:

    A.   Education and religion are powerful and influential institutions that impart values, beliefs, and knowledge considered essential to the social reproduction of individual personalities and entire cultures.

    B.   Education and religion are socializing institutions: early socialization primarily takes place in families and friendship networks; later socialization occurs in more formalized organizations created for the purposes of education and religion.

II.   EDUCATION IN HISTORICAL PERSPECTIVE

    A.   **Education** is the social institution responsible for the systematic transmission of knowledge, skills, and cultural values within a formally organized structure.

    B.   Early Formal Education

        1.   Earliest formal education probably occurred in ancient Greece and Rome where philosophers taught elite males to become thinkers and orators.

        2.   Between the fall of the Roman Empire and the beginning of the Middle Ages, only the sons of wealthy lords received formal education; other children were trained through apprenticeships in merchant and crafts guilds.

        3.   During the Middle Ages, the first colleges and universities were developed under the auspices of the church and the concept of human depravity was introduced into the curriculum.

        4.   During the Renaissance, education shifted from a focus on human depravity to the importance of developing well-rounded and liberally educated people.

        5.   With the rapid growth of industrial capitalism during the Industrial Revolution, it became necessary for workers to acquire basic skills in reading, writing, and arithmetic.

C.   Contemporary U.S. Education

      1.   In addition to teaching the basics, U.S. schools teach a myriad of topics and perform many tasks that previously were performed by other social institutions.

      2.   Controversy exists over whose values should be taught.

III.   SOCIOLOGICAL PERSPECTIVES ON EDUCATION

    A.   Functionalists view education as one of the most important components of society.

      1.   Education serves five major **manifest functions** -- open, stated, and intended goals or consequences of activities within an organization or institution:

          a.   Socialization

          b.   Transmission of culture

          c.   Social control

          d.   Social placement

          e.   Change and innovation.

      2.   Education has at least three **latent functions** -- hidden, unstated, and sometimes unintended consequences of activities within an organization or institution:

          a.   Restricting some activities

          b.   Matchmaking and production of social networks

          c.   Creation of a generation gap.

    B.   According to conflict theorists, schools perpetuate class, race, ethnic, and gender inequalities as some groups seek to maintain their privileged position at the expense of others.

      1.   Reproduction of Class: education is a vehicle for reproducing existing class relationships.

          a.   According to **Pierre Bourdieu**, children have less chance of academic success when they lack **cultural capital** -- social assets that include values, beliefs, attitudes, and competencies in language and culture.

          b.   Children from middle and upper-income families are endowed with more cultural capital than children from working-class and poverty-level families.

          c.   Class reproduction also occurs through standardized tests, ability grouping, and **tracking** -- the assignment of students to specific courses and educational programs based on their test scores, previous grades, or both.

2. The **hidden curriculum** is the transmission of cultural values and attitudes, such as conformity and obedience to authority, through implied demands found in rules, routines, and regulations of schools.

   a. Lower-class students may be disqualified from higher education and the credentials needed in a society that emphasizes **credentialism** -- a process of social selection in which class advantage and social status are linked to the possession of academic qualification.

   b. Credentialism is closely related to **meritocracy** -- a social system in which status is assumed to be acquired through individual ability and effort.

3. Unequal funding is a source of inequality in education.

   a. Most educational funds are derived from local property taxes and state legislative appropriations.

   b. Children living in affluent suburbs often attend relatively new schools and have access to the latest equipment which students in central city schools and poverty-ridden rural areas do not have.

   c. A voucher system would allow students and their families to spend a specified sum of government money to purchase education at the school of their choice.

4. Segregated and Resegregated Schools

   a. Racially segregated schools often have low retention rates, students with below-grade level reading skills, high teacher-student ratios, less-qualified teachers, and low teacher expectations

   b. Racial segregation is increasing in some U.S. schools, and efforts to bring about desegregation or integration have failed in many school districts.

   c. Even in supposedly integrated schools, tracking and ability grouping may produce resegregation at the classroom level.

   e. African American and white achievement differences increase with every year of schooling; thus, schools may reinforce, rather than eliminate, the disadvantages of race and class.

f.   The faculty of public schools does not reflect the racial-ethnic or gender makeup of the student body in many schools; Latinas/os and African Americans are underrepresented among teachers, administrators, and school board members in most systems.

g.   Regardless of their racial or ethnic group, students from poor families are 3 to 4 times more likely to become school dropouts than those from affluent families.

5.   Class, Race, and Social Reproduction in Higher Education

a.   Access to colleges and universities is determined not only by a person's prior academic record but also by the ability to pay.

b.   While Latina/o college enrollment increased from about 3.5 percent to 6.0 percent between 1976 and 1990, enrollment by African American males decreased from 9 percent to 8.4 percent, and African American women decreased from 11.5 percent to 10.6 percent. Native Americans remained at about 0.8 percent of enrollment between 1976 and 1990.

C.   Interactionist Perspective on Education

1.   Education and the Self-Fulfilling Prophecy

a.   For some students, schooling may become a **self-fulfilling prophecy** -- an unsubstantiated belief or prediction that results in behavior which makes the originally false belief come true.

b.   If a teacher (as a result of stereotypes based on the relationship between IQ and race) believes that some students of color are less capable of learning, that teacher (sometimes without even realizing it) may treat them as if they were incapable of learning.

2.   Education and Labeling

a.   IQ testing has resulted in labeling of students (e.g., African American and Mexican American children have been placed in special education classes on the basis of IQ scores when they could not understand the tests).

b.   A self-fulfilling prophecy also can result from labeling students as gifted. When some students are labeled as better than others, they

293

may achieve at a higher level because of the label, or they may face discrimination from others (e.g., Asian Americans as superintelligent).

    c.   Some analysts suggest that girls receive subtle cues from adults that lead them to attribute success to effort while boys learn to attribute success to their intelligence and ability. Conversely, girls attribute failure to lack of ability while boys attribute failure to lack of effort.

IV.   RELIGION IN HISTORICAL PERSPECTIVE
    A.   Religion and the Meaning of Life
        1.   **Religion** is a system of beliefs, symbols, and rituals, based on some sacred or supernatural realm, that guides human behavior, gives meaning to life, and unites believers into a community.
        2.   Religion seeks to answer important questions such as why we exist, why people suffer and die, and what happens when we die.
        3.   Sacred and Profane
            a.   According to Emile Durkheim, **sacred** refers to those aspects of life that are extraordinary or supernatural; those things that are set apart as "holy."
            b.   Those things people do not set apart as sacred are referred to as **profane** -- the everyday, secular or "worldly," aspects of life.
        4.   In addition to beliefs, religion also is comprised of symbols and **rituals** -- symbolic actions that represent religious meanings -- that range from songs and prayers to offerings and sacrifices.
        5.   Religions have been classified into four main categories based on their dominant belief:
            a.   **Simple supernaturalism** is the belief that supernatural forces affect people's lives either positively or negatively.
            b.   **Animism** is the belief that plants, animals, or other elements of the natural world are endowed with spirits or life forces having an impact on events in society.
            c.   **Theism** is a belief in one or more god or gods.
                (1)   **Monotheism** is a belief in a single, supreme being or god who is

responsible for significant events such as the creation of the world. Examples include: Christianity and Judaism.

        (2)   **Polytheism** is a belief in more than one god. Examples include Hinduism, Buddhism, and Shinto.

    d.   **Transcendent idealism** is a belief in sacred principles of thought and conduct, such as truth, justice, life, and tolerance for others. An example is Buddhism.

B.   Religion and Scientific Explanation

    1.   During the Industrial Revolution, rapid growth in scientific and technological knowledge gave rise to the idea that science ultimately would answer questions that previously had been in the realm of religion.

    2.   Many scholars believed that scientific knowledge would result in **secularization** -- the process by which religious beliefs, practices, and institutions lose their significance in sectors of society and culture -- but others point out a resurgence of religious beliefs and an unprecedented development of alternative religions in recent years.

V.   SOCIOLOGICAL PERSPECTIVES ON RELIGION

A.   Functionalist Perspective on Religion

    1.   According to Emile Durkheim, all religions share in common three elements: (a) beliefs held by adherents; (b) practices (rituals) engaged in collectively by believers; and (c) a moral community which results from the groups shared beliefs and practices pertaining to the sacred.

    2.   Religion has three important functions in any society:

        a.   providing meaning and purpose to life

        b.   promoting social cohesion and a sense of belonging

        c.   providing social control and support for the government.

    3.   **Civil religion** is the set of beliefs, rituals, and symbols that make sacred the values of the society and place the nation in the context of the ultimate system of meaning.

B.   Conflict Perspective on Religion

    1.   According to Karl Marx, the capitalist class uses religious ideology as a tool of domination to mislead the workers about their true interests; thus, religion is the "opiate of the people."

2. By contrast, Max Weber argued that religion could be a catalyst to produce social change.
   a. In *The Protestant Ethic and the Spirit of Capitalism*, Weber linked the teachings of **John Calvin** with the growth of capitalism.
   b. John Calvin emphasized the doctrine of predestination -- the belief that all people are divided into two groups -- the saved and the damned -- and only God knows who will go to heaven (the elect) and who will go to hell, even before they are born.
   c. Because people cannot know whether they will be saved, they look for signs that they are among the elect. As a result, people work hard, save their money and do not spend it on worldly frivolity; instead, they reinvest it in their land, equipment, and labor.
   d. As people worked ever harder to prove their religious piety, structural conditions became right in Europe for the industrial revolution, free markets, and the commercialization of the economy, which worked hand-in-hand with their religious teachings.
D. Interactionist Perspective on Religion
   1. For many people, religion serves as a reference group to help them define themselves. Religious symbols, for example, have a meaning to large bodies of people (e.g., the Star of David for Jews; the crescent moon and star for Muslims; and the cross for Christians).
   2. Her Religion and His Religion. All people do not interpret religion in the same way. Women and men may belong to the same religions, but their individual religion will not necessarily be a carbon copy of the group's entire system of beliefs.

VI. TYPES OF RELIGIOUS ORGANIZATION
   A. Some countries have an official or state religion known as an **ecclesia** -- a religious organization that is so integrated into the dominant culture that it claims as its membership all members of a society. Examples include: the Anglican Church (the official Church of England), the Lutheran Church in Sweden and Denmark, the Catholic Church in Spain, and Islam in Iran and Pakistan.
   B. The Church-Sect Typology.
      1. A **church** is a large, bureaucratically organized religious organization that tends to seek

296

accommodation with the larger society in order to maintain some degree of control over it.

   2.   Mid-way between the church and the sect is a **denomination** -- a large organized religion characterized by accommodation to society but frequently lacking in ability or intention to dominate society.

   3.   A **sect** is a relatively small religious group that has broken away from another religious organization to renew what it views as the original version of the faith.

C.   A **cult** is a religious group with practices and teachings outside the dominant cultural and religious traditions of a society

   1.   Some major religions (including Judaism, Islam, and Christianity) and some denominations (such as the Mormons) started as cults.

   2.   Cult leadership is based on charismatic characteristics of the individual, including an unusual ability to form attachments with others.

   3.   Over time, some cults undergo transformation into sects or denominations.

VII.   TRENDS IN RELIGION IN THE UNITED STATES

A.   The rise of a new fundamentalism has occurred at the same time that a number of mainline denominations have been losing membership.

B.   Some members of the political elite in Washington have vowed to bring religion "back" into schools and public life.

VIII.   EDUCATION, RELIGION, AND THE TWENTY-FIRST CENTURY

A.   Education will remain an important social institution as we enter the twenty-first century. Also remaining, however, will be the controversies over what should be taught and how to raise levels of academic achievement in the United States.

B.   Religious organizations also will continue to be important in the lives of many people; however, the influence of religious beliefs and values will be felt even by those who claim no religious beliefs of their own.

C.   In other nations, the rise of religious nationalism has led to the blending of strongly held religious and political beliefs.

D.   In the United States, the influence of religion will be evident in ongoing battles over school prayer, abortion, gay rights, and women's issues, among others. On some fronts, religion may unify people; on others, it may contribute to confrontations among individuals and groups.

ANALYZING AND UNDERSTANDING THE BOXES

After reading the chapter and studying the outline, re-read the four boxes and write down key points and possible questions for class discussion.

Sociology and Everyday Life -- "How Much Do You Know About the Impact of Religion on U.S. Education?"

Key Points:

Discussion Questions:

1.

2.

3.

Sociology and Law -- "The Separation of Church and State"

Key Points:

Discussion Questions:

1.

2.

3.

Sociology in Global Perspective -- "Religion and Women's Literacy in Developing Nations"

Key Points:

Discussion Questions:

1.

2.

3.

Sociology and Media -- "In the Media Age: The Electronic Church and the Internet"

Key Points:

Discussion Questions:

1.

2.

3.

PRACTICE TEST

MULTIPLE CHOICE QUESTIONS

Select the response that best answers the question or completes the statement:

1. _____ is the social institution responsible for the systematic transmission of knowledge, skills, and cultural values within a formally organized structure. (p. 543)
   a. Religion
   b. Mass media
   c. The government
   d. Education

2.	According to the text, all of the following statements regarding contemporary U.S. education are true, except: (p. 545)
	a.	U.S. schools today are teaching a more limited number of topics so that they can focus on the "basics."
	b.	Many parents are critical of changes that have taken place since their youth.
	c.	Recently, some parents, religious leaders, politicians, and scholars have called for the teaching of moral education in schools.
	d.	Many educators believe that their job description now encompasses numerous tasks that previously were performed by other social institutions.

3.	Which of the following is a manifest function of education? (p. 546)
	a.	Creation of a generation gap
	b.	Restricting some activities
	c.	Matchmaking and production of social networks
	d.	Social control

4.	_____ functions are hidden, unstated, and sometimes unintended consequences of activities within an organization or institution. (p. 547)
	a.	Manifest
	b.	Dormant
	c.	Latent
	d.	Covert

5.	Sociologist _____ has suggested that students come to school with differing amounts of cultural capital. (p. 548)
	a.	Emile Durkheim
	b.	Pierre Bourdieu
	c.	Jeannie Oakes
	d.	Clifford Geertz

6. The assignment of students to specific courses and educational programs based on their test scores, previous grades, or both is known as: (p. 548)
   a. tracking.
   b. the hidden curriculum.
   c. equitable assessment.
   d. class reproduction.

7. According to the _____ perspective, the hidden curriculum affects working-class and poverty-level students more than it does students from middle- and upper-income families. (p. 549)
   a. functionalist
   b. conflict
   c. interactionist
   d. feminist

8. Most educational funds are derived from: (p. 550)
   a. the federal government.
   b. private resources.
   c. local property taxes and state legislative appropriations.
   d. students' tuition and fees.

9. Which of the following statements regarding higher education is true? (p. 553)
   a. The enrollment of low-income students has increased since the 1980s.
   b. There has been an increase in scholarship funds over the past decade.
   c. College students are stratified according to their ability to pay.
   d. African American enrollment as a percentage of total college attendance has increased over the past two decades.

10. A teacher who believes (as a result of stereotypes based on the relationship between IQ and race) that some students of color are less capable of learning and treats them accordingly is an example of: (p. 554)
   a. a self-fulfilling prophecy.
   b. the labeling process.
   c. tracking.
   d. the hidden curriculum.

11. According to Emile Durkheim, _____ refers to those aspects of life that are extraordinary or supernatural. (p. 556)
    a. religion
    b. sacred
    c. profane
    d. superhuman

12. Three of the major world religions -- Christianity, Judaism, and Islam -- are characterized as: (p. 557)
    a. simple supernaturalism.
    b. animism.
    c. polytheism.
    d. monotheism.

13. According to the functionalist perspective, religion offers meaning for the human experience by: (p. 560)
    a. providing an explanation for events that create a profound sense of loss on both an individual and a group basis.
    b. offering people a reference group to help them define themselves.
    c. reinforcing existing social arrangements, especially the stratification system.
    d. encouraging the process of secularization.

14. Celebrations on Memorial Day and the Fourth of July are examples of: (p. 562)
    a. religious tolerance.
    b. civil religion.
    c. patriotic ethnocentrism.
    d. separation of church and state.

15. According to _____, the capitalist class uses religious ideology as a tool of domination. (p. 563)
    a. Emile Durkheim
    b. C. Wright Mills
    c. Karl Marx
    d. Max Weber

16. In regard to religion, Max Weber asserted that: (p. 563)
    a. church and state should be separated.
    b. religion could be a catalyst to produce social change.
    c. religion retards social change.
    d. the religious teachings of the Catholic church were directly related to the rise of capitalism.

17. The Anglican church in England and the Lutheran church in Sweden are examples of a(n): (p. 565)
    a. church.
    b. sect.
    c. denomination.
    d. ecclesia.

18. "Her religion" and "his religion" have been examined from a(n) _____ perspective. (p. 564)
    a. functionalist
    b. neo-Marxist
    c. conflict
    d. interactionist

19. In a _____, membership largely is based on birth, and children of members typically are baptized as infants. (p. 565)
    a. church
    b. sect
    c. denomination
    d. cult

20. According to the text, religious nationalism -- the blending of strongly held religious and political beliefs -- is especially strong today in: (p. 571)
    a. the United States.
    b. Middle Eastern nations.
    c. Japan.
    d. Brazil.

TRUE-FALSE QUESTIONS

T   F   1.   Debates about the appropriate relationship between public education and religion have occurred only recently in the United States. (p. 540)

T   F   2.   During the Middle Ages, the first colleges and universities were developed under the auspices of the church. (p. 543)

T   F   3.   According to Emile Durkheim, teachers are the functional equivalent of priests in teaching students about morality. (p. 546)

T   F   4.   Manifest functions in education include teaching specific subjects, such as science, history, and reading. (p. 546)

T   F   5.   Meritocracy is the process of social selection in which class advantage and social status are linked to the possession of academic qualifications. (p. 549)

T   F   6.   Current supporters of a voucher system primarily want to provide central city students with additional educational options. (p. 550)

T   F   7.   Tracking and ability grouping may produce resegregation at the classroom level. (p. 552)

T   F   8.   The issue of IQ and race/ethnicity originated with the highly controversial book by Herrnstein and Murray. (p. 554)

T   F   9.   Across cultures and in different eras, a wide variety of things have been considered sacred. (p. 556)

T   F   10.   Secularization is the process by which religious beliefs, practices, and institutions lose their significance in sectors of society and culture. (p. 559)

T   F   11.   According to functionalist theorists, religious teachings and practices help promote social cohesion by emphasizing shared symbolism. (p. 560)

T   F   12.   Karl Marx wrote *The Protestant Ethic and the Spirit of Capitalism* to explain how religion may be used by the powerful to oppress the powerless. (p. 563)

T   F   13.   Religious groups vary widely in their organizational structure. (p. 565)

T   F   14.   Denominations tend to be more tolerant and less likely than churches to expel or excommunicate members. (p. 566)

T   F   15.   "New-right" fundamentalists have encouraged secular humanism in the schools. (p. 568)

SOCIOLOGY IN OUR TIMES: DIVERSITY ISSUES

1. How has education been linked to people's resources throughout history? Do you think your own education has been influenced by your class position? If so, in what ways?

2. According to Pierre Bourdieu, what is the relationship between cultural capital and students' educational opportunities? Do you agree or disagree? Why?

3. Has there been a "hidden curriculum" in the schools you have attended? If this hidden curriculum exists, how is it related to social class and gender bias?

4. Although Box 15.3 (Sociology in Global Perspective) discusses the impact of religion on women's literacy in developing nations, do you think the same points could be made about women's experiences in the United States?

5. Analyze your own college or university based on the discussion of class, race, and social reproduction in higher education (pp. 553). Do your findings tend to confirm the assertions of conflict theorists? Why or why not?

CHAPTER FIFTEEN CROSSWORD PUZZLE

For those who enjoy crossword puzzles, here is a puzzle that contains words and names from Chapter Fifteen. Working the puzzle will help you in reviewing the chapter. The answers appear on page 309.

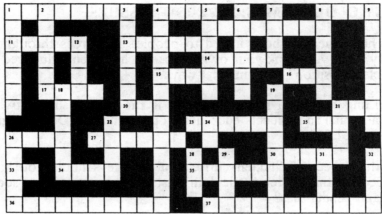

ACROSS

1. Karl Marx viewed this social institution as "[16 across] [35 across] [33 across] [16 across] masses"
4. A religious [26 down] with practices and teachings outside the dominant cultural and religious traditions of a society
8. A relatively small religious [26 down] that has broken away from another religious organization
10. A religious organization that is so integrated into the dominant culture that it claims as its membership all members of a society
11. Durkheim studied Australian aborigines and found that, to clan members, their _____ was sacred
13. A belief in a god or gods
14. Making an "A" on the Chapter 15 test would be _____ that you understand the chapter
15. In religious symbolism, females sometimes are depicted as negative, or _____ spiritual forces
16. See 1 across
17. Literacy rates for women in developing nations reflect the belief that women do not need to read or possess knowledge; these perceptions _____ to be reinforced by strong religious beliefs
20. At the start of the chapter, Jim Corbett asks if we are "going to ___ religious fundamentalists of any ilk made public policy"
21. The number of questions in Box 15.1
23. The _____ curriculum
25. Except on a test, to ___ is human, to forgive divine
26. How does the nation being first in the world in mathematics help central city students to _____ a better education?
27. Sociology of education primarily examines _____ education in industrial societies
30. See 7 down
33. See 1 across
34. In the 1960s, books considered to be racist or sexist were the subject of attacks; _____ thereafter, challenges were brought by conservative groups who found the changes objectionable
35. See 1 across
36. A belief in more than one god
37. Social system in which status is assumed to be acquired through individual ability and effort

DOWN

1. Symbolic actions that represent religious meanings
2. _____ functions: hidden, unstated, and sometimes unintended consequences
3. Animism: the belief that plants, animals, and other elements of the _____ world are endowed with spirits
4. Process of school selection in which class advantage and social status are linked to the possession of academic qualifications
5. In rural areas of some countries, a grossly inadequate school may be attached to a local _____ or mosque
6. Horace [12 down] started the free public _____ system in the U.S.
7. ____-fulfilling prophecy: an unsubstantiated belief or prediction resulting in behavior that makes the originally [30 across] belief come [21 down]
8. Refers to those aspects of [29 down] that are extraordinary or supernatural -- things that are set apart as [32 down]
9. Assignment of students to specific courses, etc., based on their previous test scores, grades, or both
12. See 6 down
18. Monotheism: a belief in a single, supreme being or god who is responsible for significant _____
19. _____ functions: open, stated, or intended consequences of activities within an organization or institution
21. See 7 down
22. Predestination is a belief that, even before they are _____, all people are divided into two groups
24. The profane ___ the everyday secular, "worldly" aspects of life
26. See 4 across
28. Functionalists assert that education provides people with an opportunity ___ upward social mobility
29. See 8 down
31. Making a bad grade can leave a _____ taste
32. See 8 down

ANSWERS TO PRACTICE TEST, CHAPTER 15

Answers to Multiple Choice Questions

1. d Education is the social institution responsible for the systematic transmission of knowledge, skills, and cultural values within a formally organized structure. (p. 543)

2. a According to the text, all of the following statements regarding contemporary U.S. education are TRUE, except: U.S. schools today are teaching a more limited number of topics so that they can focus on the "basics." (p. 545)

3. d Which of the following is a manifest function of education? social control (p. 546)

4. c Latent functions are hidden, unstated, and sometimes unintended consequences of activities within an organization or institution. (p. 547)

5. b Sociologist Pierre Bourdieu has suggested that students come to school with differing amounts of cultural capital. (p. 548)

6. a The assignment of students to specific course and educational programs based on their tests scores, previous grades, or both is known as tracking. (p. 548)

7. b According to the conflict perspective, the hidden curriculum affects working-class and poverty-level students more than it does students from middle- and upper-income families. (p. 549)

8. c Most educational funds are derived from local property taxes and state legislative appropriations. (p. 550)

9. c Which of the following statements regarding higher education is true? College students are stratified according to their ability to pay. (p. 553)

10. a A teacher who believes (as a result of stereotypes based on the relationship between IQ and race) that some students of color are less capable of learning and treats them accordingly is an example of a self-fulfilling prophecy. (p. 554)

11. b According to Emile Durkheim, sacred refers to those aspects of life that are extraordinary or supernatural. (p. 556)

12. d Three of the major world religions -- Christianity, Judaism, and Islam -- are characterized as monotheism. (p. 557)

13. a According to the functionalist perspective, religion offers meaning for the human experience by providing an explanation for events that create a profound sense of loss on both an individual and a group basis. (p. 560)

14. b Celebrations on Memorial Day and the Fourth of July are examples of civil religion. (p. 562)

15. c   According to Karl Marx, the capitalist class uses religious ideology as a tool of domination. (p. 563)

16. b   In regard to religion, Max Weber asserted that religion could be a catalyst to produce social change. (p. 563)

17. d   The Anglican church in England and the Lutheran church in Sweden are examples of an ecclesia. (p. 565)

18. d   "Her religion" and "his religion" have been examined from an interactionist perspective. (p. 564)

19. a   In a church, membership largely is based on birth, and children of members typically are baptized as infants. (p. 565)

20. b   According to the text, religious nationalism -- the blending of strongly held religious and political beliefs -- is especially strong today in Middle Eastern nations. (p. 571)

Answers to True-False Questions

1. False -- Debates about the appropriate relationship between public education and religion have occurred for many years. (p. 540)

2. True (p. 543)

3. True (p. 546)

4. True (p. 546)

5. False -- Credentialism is the process of social selection in which class advantage and social status are linked to the possession of academic qualifications. (p. 549)

6. False -- Current supporters of a voucher system primarily are members of affluent families and conservative religious groups who seek an alternative to public education and hope to use tax monies to subsidize their children's private school tuition. (p. 550)

7. True (p. 552)

8. False -- The issue of IQ and race/ethnicity first arose in regard to early-twentieth-century immigrants from Southern and Eastern Europe who scored lower, on average, on IQ tests than did Northern Europeans. (p. 554)

9. True (p. 556)

10. True (p. 559)

11. True (p. 560)

12. False -- Max Weber wrote *The Protestant Ethic and the Spirit of Capitalism* to explain how religion may be a catalyst for social change. (p. 563)

13. True (p. 565)

14. True (p. 566)

15. False -- "New-right" fundamentalists have been especially critical of secular humanism in the schools. (p. 568)

ANSWER TO CHAPTER FIFTEEN CROSSWORD PUZZLE

R	E	L	I	G	I	O	N		C	U	L	T		S		S			S	E	C	T	
I		A							R			E	C	C	L	E	S	I	A		R	
T	O	T	E	M			T	H	E	I	S	M		H		L			C		R	
U		E		A			U		D		P	R	O	O	F		O		R		K	
A		N		N			R		E	V	I	L			T	H	E		D		I	
L		T	E	N	D		A		N		E			M		A				T	E	N
S		V					L	E	T					A				T			N	G
		E			B			I		H	I	D	D	E	N		E	R	R		G	
G	A	I	N		F	O	R	M	A	L		S			I			U			H	
R		T			R			L		F		L		F	A	L	S	E			O	
O	F		S	O	O	N		I		O	P	I	A	T	E			O			O	
U								S		R		F		S				U			L	
P	O	L	Y	T	H	E	I	S	M		M	E	R	I	T	O	C	R	A	C	Y	

CHAPTER 16
POPULATION AND URBANIZATION

BRIEF CHAPTER OUTLINE
Demography: The Study of Population
    Fertility
    Mortality
    Migration
    Population Composition
Population Growth in Global Context
    The Malthusian Perspective
    The Marxist Perspective
    The Neo-Malthusian Perspective
    Demographic Transition Theory
Urbanization and the Growth of Cities
    Emergence and Evolution of the City
    Preindustrial Cities
    Industrial Cities
    Postindustrial Cities
Perspectives on Urbanization and the Growth of Cities
    Functionalist Perspectives: Ecological Models
    Conflict Perspectives: Political Economy Models
    Interactionist Perspectives: The Experience of City Life
Urban Problems in the United States
    Divided Interests: Cities, Suburbs, and Beyond
    Fiscal Crisis of the Cities
    The Crisis in Health Care
Population and Urbanization in the Twenty-First Century

CHAPTER SUMMARY
Demography is the study of the size, composition, and distribution of the population. Population growth is the result of **fertility** (births), **mortality** (deaths), and **migration**. Over two hundred years ago, Thomas Malthus warned that overpopulation would result in major global problems such as poverty and starvation. According the Marxist perspective, overpopulation occurs because of capitalists' demands for a surplus of workers to suppress wages and heighten workers' productivity. **Demographic transition** is the process by which some societies have moved from high birth and death rates to relatively low birth and death rates as a result of technological development. **Urban sociology** is the study of social relationships and political and economic structures in the city. Functionalist perspectives (ecological models) of urban growth include the concentric zone model, the

sector model, and the multiple-nuclei model. According to the political economy models of conflict theorists, urban growth is influenced by capital investment decisions, power and resource inequality, class and class conflict, and government subsidy programs. Feminist theorists suggest that cities have gender regimes; women's lives are affected by both public and private patriarchy. Interactionists focus on the positive and negative aspects of peoples' experiences in the urban settings. Urbanization, suburbanization, gentrification, and the growth of edge cities have had a dramatic impact on the U.S. population. Many central cities have experienced fiscal crises that have resulted in cuts in services, lack of maintenance of the infrastructure, and a health care crisis. Rapid global population growth is inevitable in the twenty-first century. The urban population will triple as increasing numbers of people in lesser developed and developing nations migrate from rural areas to megacities that contain a high percentage of a region's population.

LEARNING OBJECTIVES
After reading Chapter 16, you should be able to:

1. Describe the study of demography and define the basic demographic concepts.

2. Explain the Malthusian perspective on population growth.

3. Discuss the Marxist perspective on population growth and compare it with the Malthusian perspective.

4. Describe the neo-Malthusian perspective on population growth and note the significance of zero population growth to this approach.

5. Discuss demographic transition theory and explain why this theory may not apply to population growth in all societies.

6. Trace the historical development of cities and identify the major characteristics of preindustrial, industrial, and postindustrial cities.

7. Discuss functionalist perspectives on urbanization and outline the major ecological models of urban growth.

8. Compare and contrast conflict and functionalist perspectives on urban growth.

9. Describe global patterns of urbanization in core, peripheral, and semiperipheral nations.

10. Explain interactionist perspectives on urban life and note the key assumptions of the major urban theorists.

11. Discuss the major problems facing urban areas in the United States today.

KEY TERMS (defined at page number shown and in glossary)

core nations 601	megalopolis 607
crude birth rate 581	metropolis 595
crude death rate 582	migration 584
crude net migration rate586	mortality 582
demographic transition 592	peripheral nations 601
demography 579	population composition 586
emigration 584	population pyramid 587
fertility 581	semiperipheral nations 601
gentrification 597	sex ratio 586
immigration 584	succession 597
infant mortality rate 582	urban sociology 593
invasion 597	zero population growth 591

KEY PEOPLE (identified at page number shown)

CHAPTER OUTLINE

I.   DEMOGRAPHY: THE STUDY OF POPULATION

    A.   **Demography** is a subfield of sociology that examines population size, composition, and distribution.

    B.   **Fertility** is the actual level of childbearing for an individual or a population.

        1.   The **crude birth rate** is the number of live births per 1,000 people in a population in a given year.

        2.   In most areas of the world, women are having fewer children; women who have six or more children tend to live in agricultural regions where children's labor is essential to the family's economic survival and child mortality rates are very high.

    C.   A decline in **mortality** -- the incidence of death in a population -- has been the primary cause of world population growth in recent years.

        1.   The **crude death rate** is the number of deaths per 1,000 people in a population in a given year.

        2.   The **infant mortality rate** is the number of deaths of infants under 1 year of age per 1,000 live births in a given year.

    D.   **Migration** is the movement of people from one geographic area to another for the purpose of changing residency.

        1.   While **immigration** is the movement of people into a geographic area to take up residency, **emigration** is the movement of people out of a geographic area to take up residency elsewhere.

        2.   The **crude net migration rate** is the net number of migrants (total in-migrants minus total out-migrants) per 1,000 people in a population in a given year.

    E.   **Population composition** is the biological and social characteristics of a population, including age, sex, race, marital status, education, occupation, and income.

1. The **sex ratio** is the number of males for every hundred females in a given population; a sex ratio of 100 indicates an equal number of males and females.
2. A **population pyramid** is a graphic representation of the distribution of a population by sex and age.

II. POPULATION GROWTH IN GLOBAL CONTEXT
   A. The Malthusian Perspective
      1. According to **Thomas Robert Malthus**, the population (if left unchecked) would exceed the available food supply; population would increase in a geometric progression (2, 4, 8, 16 . . .) while the food supply would increase only by an arithmetic progression (1, 2, 3, 4 . . .).
      3. This situation could end population growth and perhaps the entire population unless positive checks (such as famines, disease and wars) or preventive checks (such as sexual abstinence and postponement of marriage) intervened.
   B. The Marxist Perspective
      1. According to Karl Marx and Friedrich Engels, food supply does not have to be threatened by overpopulation; through technology, food for a growing population can be produced.
      2. Overpopulation occurs because capitalists want a surplus of workers (an industrial reserve army) to suppress wages and force employees to be more productive.
      3. Overpopulation will lead to the eventual destruction of capitalism; when workers become dissatisfied, they will develop class consciousness because of shared oppression.
   C. The Neo-Malthusian Perspective
      1. Neo-Malthusians (or "New Malthusians") reemphasized the dangers of overpopulation and suggested that an exponential growth pattern is occurring.
      2. Overpopulation and rapid population growth result in global environmental problems, and people should be encouraging **zero population growth** -- the point at which no population increase occurs from year to year because the number of births plus immigrants is equal to the number of deaths plus emigrants.

D. Demographic Transition Theory
   1. **Demographic transition** is the process by which some societies have moved from high birth and death rates to relatively low birth and death rates as a result of technological development.
   2. Demographic transition is linked to four stages of economic development:
      a. Stage 1: Preindustrial Societies -- little population growth occurs, high birth rates are offset by high death rates.
      b. Stage 2: Early Industrialization -- significant population growth occurs, birth rates are relatively high while death rates decline.
      c. Stage 3: Advanced Industrialization and Urbanization -- very little population growth occurs, both birth rates and death rates are low.
      d. Stage 4: Postindustrialization -- birth rates continue to decline as more women are employed full-time and raising children becomes more costly; population growth occurs slowly, if at all, due to a decrease in the birth rate and a stable death rate.
   3. Critics suggest that demographic transition theory may not accurately explain population growth in all societies; this theory may best explains growth in Western societies.

III. URBANIZATION AND THE GROWTH OF CITIES
   A. **Urban sociology** is a subfield of sociology that examines social relationships and political and economic structures in the city.
   B. Emergence and Evolution of the City
      1. Cities are a relatively recent innovation as compared with the length of human existence. According to Gideon Sjoberg, three preconditions must be present in order for a city to develop:
         a. A favorable physical environment
         b. An advanced technology that could produce a social surplus
         c. A well-developed political system to provide social stability to the economic system.
      2. Sjoberg places the first cities in the Mesopotamian region or areas immediately adjacent to it at about 3500 B.C.E; however, not all scholars agree on this point.

315

C.   Preindustrial Cities
    1.   The largest preindustrial city was Rome.
    2.   Preindustrial cities were limited in size because of crowded housing conditions, lack of adequate sewage facilities, limited food supplies, and lack of transportation to reach the city.
    3.   Many preindustrial cities had a sense of community -- a set of social relationships operating within given spatial boundaries that provide people with a sense of identity and a feeling of belonging.

D.   Industrial Cities
    1.   The nature of the city changed as factories arose and new forms of transportation and agricultural production made it easier to leave the countryside and move to the city.
    2.   New York City became the first U.S. **metropolis** -- one or more central cities and their surrounding suburbs that dominate the economic and cultural life of a region.
    3.   People lived in close proximity to factories so that they could walk to work; many lived in overcrowded conditions that lacked sanitation and a clean water supply.

E.   Postindustrial Cities
    1.   Since the 1950s, postindustrial cities have emerged as the U.S. economy has gradually shifted from secondary (manufacturing) to tertiary (service and information processing) production.
    2.   Postindustrial cities are dominated by "light" industry, such as computer software manufacturing, information-processing services, educational complexes, medical centers, retail trade centers, and shopping malls.

IV.   PERSPECTIVES ON URBANIZATION AND THE GROWTH OF CITIES
    A.   Functionalist Perspectives: Ecological Models
        1.   **Robert Park** based his analysis of the city on human ecology -- the study of the relationship between people and their physical environment -- and found that economic competition produces certain regularities in land-use patterns and population distributions.

2. Concentric zone model
   a. Based on Park's ideas, **Ernest W. Burgess** developed a model that views the city as a series of circular zones, each characterized by a different type of land use, that developed from a central core: (1) the central business district and cultural center; (2) the zone of transition -- houses where wealthy families previously lived that have now been subdivided and rented to persons with low incomes; (3) working-class residences and shops, and ethnic enclaves; (4) homes for affluent families, single-family residences of white-collar workers, and shopping centers; and (5) a ring of small cities and towns comprised of estates owned by the wealthy and houses of commuters who work in the city.
   b. Two important ecological processes occur: **invasion** is the process by which a new category of people or type of land use arrives in an area previously occupied by another group or land use, and **succession** is the process by which a new category of people or type of land use gradually predominates in an area formerly dominated by another group or activity.
   c. **Gentrification** is the process by which members of the middle and upper-middle classes, especially whites, move into the central city area and renovate existing properties.
3. Sector Model
   a. **Homer Hoyt**'s sector model emphasizes the significance of terrain and the importance of transportation routes in the layout of cities.
   b. Residences of a particular type and value tend to grow outward from the center of the city in wedge-shaped sectors with the more expensive residential neighborhoods located along the higher ground near lakes and rivers, or along certain streets that stretch from the downtown area.
   c. Industrial areas are located along river valleys and railroad lines; middle class residences exist

317

on either side of wealthier neighborhoods; lower class residential areas border the central business area and the industrial areas.

4. Multiple-Nuclei Model
   a. According to **Chauncey Harris** and **Edward Ullman**, cities have numerous centers of development; as cities grow, they annex outlying townships.
   b. In addition to the central business district, other nuclei develop around activities such as an educational institution or a medical complex; residential neighborhoods may exist close to or far away from these nuclei.

5. Contemporary Urban Ecology.
   a. Amos Hawley viewed urban areas as complex social systems in which growth patterns are based on advances in transportation and communication.
   b. Social area analysis examines urban populations in terms of economic status, family status, and ethnic classification.

B. Conflict Perspectives: Political Economy Models
   1. According to Marx, cities are arenas in which the intertwined processes of class conflict and capital accumulation take place; class consciousness is more likely to occur in cities where workers are concentrated.
   2. Three major themes are found in political economy models:
      a. Patterns of urban growth and decline are affected by: (1) economic factors such as capitalist investments; and (2) political factors, including governmental protection of private property and promotion of the interests of business elites and large corporations.
      b. Urban space has both an exchange value and a use value: (1) exchange value refers to the profits industrialists, developers, and bankers make from buying, selling, and developing land and buildings; (2) use value is the utility of space, land, and buildings for family life and neighborhood life.
      c. Structure and agency are both important in understanding how urban development takes place: (1) structure refers to institutions such

318

as state bureaucracies and capital investment circuits that are involved in the urban development process; and (2) agency refers to human actors who participate in land use decisions, including developers, business elites, and activists protesting development.

3. According to political economy models, urban growth is influenced by capital investment decisions, power and resource inequality, class and class conflict, and government subsidy programs.

4. Gender Regimes in Cities

   a. According to feminist perspectives, urbanization reflects the workings of the political economy and patriarchy.

   b. Different cities have different gender regimes -- prevailing ideologies of how women and men should think, feel, and act; how access to positions and control of resources should be managed; and how women and men should relate to each other.

   c. Gender intersects with class and race as a form of oppression, especially for lower-income women of color who live in central cities.

5. Global Patterns

   a. Urbanization differs widely based on a nation's level of development: (1) **core nations** are dominant capitalist centers characterized by high levels of industrialization and urbanization; (2) **peripheral nations** are dependent on core nations for capital, have little or no industrialization, and have uneven patterns of urbanization; and (3) **semiperipheral nations** are more developed than the peripheral nations but less developed than core nations.

   b. Uneven economic growth results from capital investment by core nations; disparity between the rich and the poor within these nations is increased in the process.

   c. Thousands of low-wage workers who moved to urban areas seeking work have settled in shantytowns on the edge of the city or in low-cost rental housing in central city slums. Squatters, who occupy land without any legal title to it, are the most rapidly growing segment of the population in many global cities.

C.   Interactionist Perspectives: The Experience of City Life

    1.   Simmel's View of City Life

        a.   According to **Georg Simmel**, urban life is highly stimulating; it shapes people's thoughts and actions.

        b.   However, many urban residents avoid emotional involvement with each other and try to ignore events taking place around them.

        c.   City life is not completely negative; urban living can be liberating -- people have opportunities for individualism and autonomy.

    2.   Urbanism as a Way of Life

        a.   **Louis Wirth** suggested that urbanism is a "way of life." Urbanism refers to the distinctive social and psychological patterns of city life.

        b.   Size, density, and heterogeneity result in an elaborate division of labor and in spatial segregation of people by race/ethnicity, class, religion, and/or lifestyle; a sense of community is replaced by the "mass society" -- a large-scale, highly institutionalized society in which individuality is supplanted by mass media, faceless bureaucrats, and corporate interests.

    3.   Gans's Urban Villagers

        a.   According to **Herbert Gans,** not everyone experiences the city in the same way; some people develop strong loyalties and a sense of community within central city areas that outsiders may view negatively.

        b.   Five major categories of urban dwellers are: (1) cosmopolites -- students, artists, writers, musicians, entertainers, and professionals who choose to live in the city because they want to be close to its cultural facilities; (2) unmarried people and childless couples who live in the city because they want to be close to work and entertainment; (3) ethnic villagers who live in ethnically segregated neighborhoods; (4) the deprived -- individuals who are very poor and see few future prospects; and (5) the trapped -- those who cannot escape the city, including downwardly mobile persons, older persons, and persons with addictions.

    4.   Gender and City Life

        a.   According to **Elizabeth Wilson**, some men view the city as sexual space in which women, based on their sexual desirability and accessibility, are categorized as prostitutes, lesbians, temptresses, or virtuous women in need of protection.

        b.   More affluent, dominant group women are more likely to be viewed as virtuous women in need of protection while others are placed in less desirable categories.

V.   URBAN PROBLEMS IN THE UNITED STATES

    A.   Divided Interests: Cities, Suburbs, and Beyond

        1.   Since World War II, the U.S. population has shifted dramatically as many people have moved to the suburbs.

        2.   Suburbanites rely on urban centers for employment and some services but pay property taxes to suburban governments and school districts; some affluent suburbs have state-of-the-art school districts and infrastructure while central city services and school districts lack funds.

        3.   Race, Class, and Suburbs

            a.   The intertwining impact of race and class is visible in the division between central cities and suburbs.

            b.   Most suburbs are predominantly white; many upper-middle and upper-class suburbs remain virtually all white; people of color who live in suburbs often are resegregated.

        4.   Beyond the Suburbs

            a.   Edge cities initially develop as residential areas beyond central cities and suburbs; then retail establishments and office parks move into the area and create an unincorporated edge city.

            b.   Corporations move to edge cities because of cheaper land and lower utility rates and property taxes.

        5.   Likewise, Sunbelt cities grew in the 1970s, as millions moved from the north and northeastern states to southern and western states where there were more jobs and higher wages, lower taxes, pork-barrel programs funded by federal money that created jobs

and encouraged industry, and the presence of high technology industries.

B. Fiscal Crisis of the Cities
1. During the 1980s, many cities experienced fiscal crises as federal and state aid was drastically cut.
2. Services had to be cut or taxes raised at the same time that the tax base was shrinking due to suburban flight.

C. The Crisis in Health Care
1. Although health care is a problem throughout the United States, it especially is problematic in large urban areas where there are higher rates of poverty.
2. Poor people are more likely to have certain types of diseases (e.g., tuberculosis) and problems associated with lack of preventive care (infants with low birth weights).

VI. POPULATION AND URBANIZATION IN THE TWENTY-FIRST CENTURY
A. Rapid global population growth is inevitable: although death rates have declined in many developing nations, there has not been a corresponding decrease in birth rates.
B. In the future, many people will live in a **megalopolis** -- a continuous concentration of two or more cities and their suburbs that form an interconnected urban area.
C. At the macrolevel, we can do little about population and urbanization; at the microlevel, we may be able to exercise some degree of control over our communities and our own lives.

ANALYZING AND UNDERSTANDING THE BOXES
After reading the chapter and studying the outline, re-read the four boxes and write down key points and possible questions for class discussion.

Sociology and Everyday Life -- "How Much Do You Know About HIV/AIDS in Global-Human Perspective?"

Key Points:

Discussion Questions:

1.

2.

3.

Sociology and Law -- "HIV Testing as a Criteria for Immigration"

Key Points:

Discussion Questions:

1.

2.

3.

Sociology in Global Perspective -- "The AIDS Epidemic in Africa"

Key Points:

Discussion Questions:

1.

2.

3.

Sociology and Media -- "AIDS in the News"

Key Points:

Discussion Questions:

1.

2.

3.

PRACTICE TEST

MULTIPLE CHOICE QUESTIONS

Select the response that best answers the question or completes the statement:

1. Demography is a subfield of sociology that examines: (p. 579)
    a.   population size.
    b.   population composition.
    c.   population distribution.
    d.   all of the above.

2. _____ is the actual level of childbearing for an individual or a population, while _____ is the potential number of children that could be born if every woman reproduced at her maximum biological capacity. (p. 581)
    a.   Birth rate -- fertility rate
    b.   Fertility rate -- birth rate
    c.   Fertility -- fecundity
    d.   Fecundity -- fertility

3. In 1993, the leading cause of death in the United States was: (p. 583)
    a.   influenza/pneumonia.
    b.   heart disease.
    c.   kidney disease.
    d.   HIV.

4. The average lifetime in years of people born in a specific year is known as: (p. 584)
    a.   life expectancy.
    b.   the crude mortality rate.
    c.   the longevity table.
    d.   age-specific death rates.

5. _____ is the movement of people out of a geographic area to take up residency elsewhere. (p. 584)
    a.   Immigration
    b.   Emigration
    c.   Transmigration
    d.   Ex-migration

6. According to Thomas Malthus's perspective on population: (p. 587)
   a. the population would increase in a geometric progression while the population would increase in an arithmetic progression.
   b. the population would increase in an arithmetic progression while the population would increase in a geometric progression.
   c. the food supply is not threatened by overpopulation because technology makes it possible to produce the food and other goods needed to meet the demands of a growing population.
   d. societies move through a process of demographic transition.

7. According to Karl Marx and Friedrich Engels's perspective on population: (p. 587)
   a. the population would increase in a geometric progression while the food supply would increase in an arithmetic progression.
   b. the population would increase in an arithmetic progression while the food supply would increase in a geometric progression.
   c. the food supply is not threatened by overpopulation because technology makes it possible to produce the food and other goods needed to meet the demands of a growing population.
   d. societies move through a process of demographic transition.

8. According to the demographic transition theory, significant population growth occurs because birth rates are relatively high while death rates decline in the _____ stage of economic development. (p. 592)
   a. preindustrial
   b. early industrial
   c. advanced industrial
   d. postindustrial

9. All of the following statements regarding preindustrial cities are true, except: (p. 594)
   a. the largest preindustrial city was Rome.
   b. preindustrial cities were limited in size because of crowded housing conditions and a lack of adequate sewage facilities.
   c. food supplies were limited in preindustrial cities.
   d. preindustrial cities lacked a sense of community.

10. The *Gemeinschaft* and *Gesellschaft* typology originated with: (p. 594)
    a. Emile Durkheim.
    b. Gideon Sjoberg.
    c. Max Weber.
    d. Ferdinand Tonnies.

11. _____ refers to one or more central cities and their surrounding suburbs that dominate the economic and cultural life of a region. (p. 595)
    a. Urban sprawl
    b. Megalopolis
    c. Metropolis
    d. Urbanization

12. Postindustrial cities are characterized by: (p. 595)
    a. "light" industry, information-processing services, educational complexes, retail trade centers, and shopping malls.
    b. the growth of the factory system.
    c. agricultural production.
    d. "heavy" industry, such as automobile manufacturing.

13. Ecological models of urban growth are based on a _____ perspective. (p. 596)
    a. functionalist
    b. conflict
    c. neo-Marxist
    d. interactionist

14. All of the following are ecological models of urban growth, except the: (p. 597)
    a. concentric zone model.
    b. sector model.
    c. urban sprawl model.
    d. multiple nuclei model.

15. An upper-middle class doctor who moves her family from the suburbs into the central city to renovate an older home is an example of: (p. 597)
    a. succession.
    b. gentrification.
    c. ex-suburbanization.
    d. downward immigration.

16. According to political economy models, urban growth is: (p. 599)
    a. influenced by terrain and transportation.
    b. based on the clustering of people who share similar characteristics.
    c. linked with peaks and valleys in the economic cycle.
    d. influenced by capital investment decisions, power and resource inequality, and government subsidy programs.

17. _____ refers to the tendency of some neighborhoods, cities, or regions to grow and prosper while others stagnate and decline. (p. 600)
    a. Invasion
    b. Succession
    c. Gentrification
    d. Uneven development

18. The United States, Japan, and Germany are _____ nations. (p. 601)
    a. core
    b. peripheral
    c. semiperipheral
    d. globalized

19. Sociologist _____ has argued that urbanism is a "way of life." (p. 602)
    a. Herbert Gans
    b. Georg Simmel
    c. Louis Wirth
    d. Elizabeth Wilson

20. According to sociologists, edge cities: (p. 606)
    a. initially develop as industrial parks.
    b. drain taxes from central cities and older suburbs.
    c. have existed since World War II.
    d. always correspond to municipal boundaries.

TRUE-FALSE QUESTIONS

T   F   1. According to the text, HIV/AIDS is one of the most significant global-human problems we face. (p. 577)

T   F   2. The world's population is increasing by 94 million people per year. (p. 579)

T   F   3. In most areas of the world, women are having more children. (p. 581)

T   F.   4.   The primary cause of world population growth in recent years has been a decline in mortality. (p. 582)

T   F   5.   Pull factors of migration include political unrest and war. (p. 585)

T   F   6.   Mortality is more difficult to measure than fertility and migration. (p. 586)

T   F   7.   Population pyramids are graphic representations of the distribution of a population by sex and age. (p. 587)

T   F   8.   According to the Marxist perspective, overpopulation occurs because capitalists desire to have a surplus of workers so as to suppress wages and increase workers' productivity. (p. 589)

T   F   9.   About 75 percent of the U.S. population lives in cities. (p. 593)

T   F   10.   The sector model views the city as a series of circular areas or zones, each characterized by a different type of land use, that developed from a central core. (p. 596)

T   F   11.   According to conflict theorists, cities grow and decline by chance. (p. 599)

T   F   12.   According to feminist theorists, public patriarchy may be perpetuated by cities through policies that limit women's access to paid work and public transportation. (p. 600)

T   F   13.   Interactionists examine the experience of urban life rather than the political economy of the city. (p. 602)

T   F   14.   Herbert Gans has suggested that city dwellers live in urban areas by choice. (p. 603)

T   F   15.   Nationally, most suburbs are predominantly white. (p. 605)

SOCIOLOGY IN OUR TIMES: DIVERSITY ISSUES

1.   Why is HIV/AIDS a "global-human" problem? Does HIV/AIDS "discriminate" by race, class, gender, or age? Why or why not?

2.   Is the community or city in which you live the "same" for men and women? Are women and men equally "safe" on the streets? Where do you feel most secure? Least secure? Does gender interact with class and race as a form of oppression for lower-income women of color? (pp. 600, 604)

3.   How are race and class intertwined with the contemporary problems of cities and suburbs? Are people in your community or city segregated, either voluntarily or involuntarily, by race and class?

CHAPTER SIXTEEN CROSSWORD PUZZLE

For those who enjoy crossword puzzles, here is a puzzle that contains words and names from Chapter Sixteen. Working the puzzle will help you in reviewing the chapter. The answers appear on page 333.

ACROSS

1. The movement of people into a geographic area to take up residency
6. The distinctive social and psychological patterns of life typically found in the city
9. The actual level of childbearing for an individual or a population
10. See 1 down
12. According to sociologist _____ Simmel, urban life is highly stimulating, and it shapes people's thoughts and actions
14. According to Herbert Gans, _____ villagers live in segregated neighborhoods
15. A sex ratio of 100 indicates a _____ number of males and females in the population
17. Rae Lewis-Thornton is quoted in the text as saying, "I am dying because I had one sexual partner too many. And I'm here to _____ you now one is all it takes."
18. From the list of megalopolises in the world, don't _____ Rio de Janeiro
19. Gentrification: the process by which members of the middle and upper-middle classes move into the central city -- and renovate existing structures
21. See 3 down
24. _____ are getting to the end of the text and of the course: congratulations!
25. Homer Hoyt developed the _____ model of urban growth
26. Initials of first U.S. metropolis
27. See 21 down
28. The movement of people out of a geographic area to take up residency elsewhere
29. Sex _____: the number of males for every hundred females in a given population

DOWN

1. _____ mortality rate: the number of deaths of children under [10 across] year of age per 1,000 live births in a given year
2. The incidence of death in a population
3. Most people who live in a [21 across] enjoy life there, even though it is not hard to find something to _____ about
4. Thomas Malthus [7 down] that the population would increase in a geometric progression while the food supply would only increase by an _____ progression
5. _____-Malthusians have reemphasized the dangers of over population
6. Invasion: the process by which a new category of people or type of land _____ arrives in an area previously occupied by another group or type of land _____
7. See 4 down
8. According to the text, Mexico City and Buenos Aires are examples
11. [20 down] _____: dominant capitalist centers characterized by a high level of industrialization and urbanization
16. One example of a dominant capitalist center is the _____. _____.
20. See 11 down
21. _____ birth [27 across]: the number of live births per 1,000 people in a population in a given year
22. Within nations, people from large cities may be pulled to rural areas by lower crime rates, while _____ people from small towns may be drawn to cities by better jobs
23. The population composition characteristics that are examined include age, sex, race, and _____ characteristics

ANSWERS TO PRACTICE TEST, CHAPTER 16

Answers to Multiple Choice Questions

1.   d   Demography is a subfield of sociology that examines population size, population composition, and population distribution. Thus "all of the above" is the best answer. (p. 579)

2.   c   Fertility is the actual level of childbearing for an individual or a population, while fecundity is the potential number of children that could be born if every woman reproduced at her maximum biological capacity. (p. 581)

3.   b   In 1993, the leading cause of death in the United States was heart disease. (p. 583)

4.   a   The average lifetime in years of people born in a specific year is known as life expectancy. (p. 584)

5.   b   Emigration is the movement of people out of a geographic area to take up residency elsewhere. (p. 584)

6.   a   According to Thomas Malthus's perspective on population, the population would increase in an geometric progression while the food supply would increase in a arithmetic progression. (p. 587)

7.   c   According to Karl Marx and Friedrich Engels's perspective on population, the food supply is not threatened by overpopulation because technology makes it possible to produce the food and other goods needed to meet the demands of a growing population. (p. 587)

8.   b   According to the demographic transition theory, significant population growth occurs because birth rates are relatively high while death rates decline in the early industrial stage of economic development. (p. 592)

9.   d   All of the following statements regarding preindustrial cities are true, <u>except</u> preindustrial cities lacked a sense of community. (p. 594)

10.   d   The *Gemeinschaft* and *Gesellschaft* typology originated with Ferdinand Tonnies. (p. 594)

11.   c   Metropolis refers to one or more central cities and their surrounding suburbs that dominate the economic and cultural life of a region. (p. 595)

12.   a   Postindustrial cities are characterized by "light" industry, information-processing services, educational complexes, retail trade centers, and shopping malls. (p. 595)

13.   a   Ecological models of urban growth are based on a functionalist perspective. (p. 596)

14. c   All of the following are ecological models of urban growth, except the urban sprawl model. (p. 597)
15. b   An upper-middle class doctor who moves her family from the suburbs into the central city to renovate an older home is an example of gentrification. (p. 597)
16. d   According to political economy models, urban growth is influenced by capital investment decisions, power and resource inequality, and government subsidy programs. (p. 599)
17. d   Uneven development refers to the tendency of some neighborhoods, cities, or regions to grow and prosper while others stagnate and decline. (p. 600)
18. a   The United States, Japan, and Germany are core nations. (p. 601)
19. c   Sociologist Louis Wirth has argued that urbanism is a "way of life." (p. 602)
20. b   According to sociologists, edge cities drain taxes from central cities and older suburbs. (p. 606)

Answers to True-False Questions

1.   True (p. 577)
2.   True (p. 579)
3.   False -- In most areas of the world, women are having fewer children. (p. 581)
4.   True (p. 582)
5.   False. Push factors of migration include political unrest and war. (p. 585)
6.   False -- Mortality is more difficult to measure than fertility and migration. (p. 586)
7.   True (p. 587)
8.   True (p. 589)
9.   True (p. 593)
10.   False -- The concentric zone model views the city as a series of circular areas or zones, each characterized by a different type of land use, that developed from a central core. (p. 596)
11.   False -- According to conflict theorists, cities do not grow and decline by chance. Rather, they are the product of specific decisions made by members of the capitalist class and political elites. (p. 599)
12.   True (p. 600)
13.   True (p. 602)
14.   False -- Although Herbert Gans has suggested that some city dwellers (especially cosmopolites and unmarried people and childless couples who want to live close to work and entertainment) do live in the cities

by choice, others (including the deprived and the trapped) can find no escape from the city. (p. 603)

15. True (p. 605)

ANSWER TO CHAPTER SIXTEEN CROSSWORD PUZZLE

```
I M M I G R A T I O N   U R B A N I S M
N   O   R   R       E   S   R       E
F E R T I L I T Y   O N E   G E O R G
A   T   P   T     E   A     U       A
N   A   E T H N I C   T     E Q U A L
T E L L   M     O M I T     D   S   O
    I   A R E A   L   O       C     P
C I T Y     T     O   N     O   O   O
R   Y O U   I     G   S E C T O R   L
U     U     C   N Y C       H   E   I
D     N           R A T E           S
E M I G R A T I O N       R A T I O
```

CHAPTER 17
COLLECTIVE BEHAVIOR AND SOCIAL CHANGE

BRIEF CHAPTER OUTLINE
Collective Behavior
    Conditions for Collective Behavior
    Dynamics of Collective Behavior
    Distinctions Regarding Collective Behavior
    Types of Crowd Behavior
    Explanations of Crowd Behavior
    Mass Behavior
Social Movements
    Types of Social Movements
    Causes of Social Movements
    Stages in Social Movements
Social Change: Moving into the Twenty-First Century
    The Physical Environment and Change
    Population and Change
    Technology and Change
    Social Institutions and Change
    A Few Final Thoughts

CHAPTER SUMMARY
Social change is the alteration, modification, or transformation of public policy, culture, or social institutions over time. Such change usually is brought about by **collective behavior** -- voluntary, often spontaneous activity that is engaged in by a large number of people and typically violates dominant group norms and values. A **crowd** is a relatively large number of people who are in one another's immediate vicinity. Five categories of crowds have been identified: (1) casual crowds are relatively large gatherings of people who happen to be in the same place at the same time; (2) conventional crowds are comprised of people who specifically come together for a scheduled event and thus share a common focus; (3) expressive crowds provide opportunities for the expression of some strong emotion; (4) acting crowds are collectivities so intensely focused on a specific purpose or object that they may erupt into violent or destructive behavior; and (5) protest crowds are gatherings of people who engage in activities intended to achieve specific political goals. Protest crowds sometime participate in **civil disobedience** -- nonviolent action that seeks to change a policy or law by refusing to comply with it. Explanations of crowd behavior include contagion theory, social unrest and circular reaction, convergence theory, and emergent norm theory. Examples of **mass behavior** -- collective behavior that takes place when people respond to the same event in much the same way -- include rumors, gossip, mass hysteria, fads, fashions, and public opinion. The major types of **social movements** -- organized groups that act consciously to promote or resist change through collective action -- are reform movements, revolutionary movements, religious movements, alternative movements, and resistance movements. Sociological theories explaining social movements include relative deprivation theory, value-added theory, resource mobilization theory,

and recent emerging perspectives. Social change produces many challenges that remain to be resolved: environmental problems, changes in the demographics of the population, and new technology that benefits some -- but not all -- people. As we head into the twenty-first century, we must use our sociological imaginations to help resolve these problems.

LEARNING OBJECTIVES
After reading Chapter 17, you should be able to:

1. Define collective behavior and describe the conditions necessary for such behavior to occur.

2. Distinguish between crowds and masses and identify casual, conventional, expressive, acting, and protest crowds.

3. Discuss the key elements of these four explanations of collective behavior: contagion theory, social unrest and circular reaction, convergence theory, and emergent norm theory.

4. Define mass behavior and describe the most frequent types of this behavior.

5. Describe social movements and note when and where they are most likely to develop.

6. Differentiate among the five major types of social movements based on their goals and the amount of change they seek to produce.

7. Compare relative deprivation theory and value-added theory as explanations of why people join social movements.

8. State the key assumptions of resource mobilization theory.

9. Describe "new social movements" and give examples of issues of concern to participants in these movements.

10. Identify the stages in social movements and explain why social movements may be an important source of social change.

11. Describe the effects of physical environment, population trends, technological development, and social institutions on social change.

KEY TERMS (defined at page number shown and in glossary)

civil disobedience 623	panic 623
collective behavior 617	propaganda 630
crowd 620	public opinion 629
gossip 627	riot 621
mass 620	rumors 627
mass behavior 626	social change 616
mass hysteria 628	social movement 630
mob 621	terrorism 633

KEY PEOPLE (identified at page number shown)

Herbert Blumer 629	Clark McPhail and
Pierre Bourdieu 629	Ronald T. Wohlstein 623
Lory Britt 631	William Ogburn 644
Stella M. Capek 639	Robert E. Park 625
Steven E. Clayman 626	Alan Scott 638
Riley E. Dunlap 630	Georg Simmel 629
Kai Erikson 631	Neal Smelser 635
William A. Gamson 620, 637	Charles Tilly 637
J. Stephen Kroll-Smith and	Ralph H. Turner and
Stephen Robert Couch 637	Lewis M. Killian 620, 626
Gustave Le Bon 624	Thorstein Veblen 629
John Lofland 620	

CHAPTER OUTLINE

I. COLLECTIVE BEHAVIOR
   A. **Social change** is the alteration, modification, or transformation of public policy, culture, or social institutions over time; such change usually is brought about by **collective behavior** -- relatively spontaneous, unstructured activity that typically violates established social norms.
   B. Conditions for Collective Behavior
      1. Collective behavior occurs as a result of some common influence or stimuli which produces a response from a collectivity -- a relatively large number of people who mutually transcend, bypass, or subvert established institutional patterns and structures.
      2. Major factors that contribute to the likelihood that collective behavior will occur are:

　　　　　　　　a.　　Structural factors that increase the chances of people responding in a particular way;
　　　　　　　　b.　　Timing;
　　　　　　　　c.　　A breakdown in social control mechanisms and a corresponding feeling of normlessness;
　　　　　　　　d.　　A common stimulus.
C.　　Dynamics of Collective Behavior
　　　　1.　　People may engage in collective behavior when they find that their problems are not being solved through official channels; as the problem appears to grow worse, organizational responses become more defensive and obscure.
　　　　2.　　People's attitudes are not always reflected in their political and social behavior; free riders are people who enjoy the benefits produced by a group even though they have not helped support it.
　　　　3.　　People act collectively in ways they would not act singly due to:
　　　　　　　　a.　　The noise and activity around them.
　　　　　　　　b.　　A belief that it is the only way to fight those with greater power and resources.
D.　　Distinctions Regarding Collective Behavior
　　　　1.　　People engaging in collective behavior may be a:
　　　　　　　　a.　　**Crowd** -- a relatively large number of people who are in one another's immediate face-to-face presence; or
　　　　　　　　b.　　**Mass** -- a number of people who share an interest in a specific idea or issue but who are not in one another's immediate physical vicinity.
　　　　2.　　Collective behavior also may be distinguished by the dominant emotion expressed (e.g., fear, hostility, joy, grief, disgust, surprise, or shame).
E.　　Types of Crowd Behavior
　　　　1.　　**Herbert Blumer** divided crowds into four categories:
　　　　　　　　a.　　Casual crowds -- relatively large gatherings of people who happen to be in the same place at the same time; if they interact at all, it is only briefly.
　　　　　　　　b.　　Conventional crowds -- people who specifically come together for a scheduled event and thus share a common focus.
　　　　　　　　c.　　Expressive crowds -- people releasing their pent-up emotions in conjunction with others who experience similar emotions.
　　　　　　　　d.　　Acting crowds -- collectivities so intensely focused on a specific purpose or object that they may erupt into violent or destructive behavior. Examples:
　　　　　　　　　　　(1)　　A **mob** -- a highly emotional crowd whose members engage in, or are ready to engage in, violence against a

337

specific target which may be a person, a category of people, or physical property.

    (2)   A **riot** -- violent crowd behavior fueled by deep-seated emotions but not directed at a specific target.

    (3)   A **panic** -- a form of crowd behavior that occurs when a large number of people react with strong emotions and self-destructive behavior to a real or perceived threat.

2.   To these four types of crowds, Clark McPhail and Ronald T. Wohlstein added protest crowds -- crowds that engage in activities intended to achieve specific political goals.

    a.   Protest crowds sometimes take the form of **civil disobedience** -- nonviolent action that seeks to change a policy or law by refusing to comply with it.

    b.   At the grassroots level, protests often are seen as the only way to call attention to problems or demand social change.

F.   Explanations of Crowd Behavior

1.   According to contagion theory, people are more likely to engage in antisocial behavior in a crowd because they are anonymous and feel invulnerable; Gustave Le Bon argued that feelings of fear and hate are contagious in crowds because people experience a decline in personal responsibility.

2.   According to **Robert Park**, social unrest is transmitted by a process of circular reaction -- the interactive communication between persons in such a way that the discontent of one person is communicated to another who, in turn, reflects the discontent back to the first person.

3.   Convergence theory focuses on the shared emotions, goals, and beliefs many people bring to crowd behavior.

    a.   From this perspective, people with similar attributes find a collectivity of like-minded persons with whom they can release their underlying personal tendencies.

    b.   Although people may reveal their "true selves" in crowds, their behavior is not irrational; it is highly predictable to those who share similar emotions or beliefs.

4.   According to Ralph Turner's and Lewis Killian's emergent norm theory, crowds develop their own definition of the situation and establish norms for behavior that fits the occasion.

<ol type="a">
<li>Emergent norms occur when people define a new situation as highly unusual or see a long-standing situation in a new light.</li>
<li>Emergent norm theory points out that crowds are not irrational; new norms are developed in a rational way to fit the needs of the immediate situation.</li>
</ol>

G. Mass Behavior

1. **Mass behavior** is collective behavior that takes place when people (who often are geographically separated from one anther) respond to the same event in much the same way. The most frequent types of mass behavior are:

2. **Rumors** -- unsubstantiated reports on an issue or subject -- and **gossip** -- rumors about the personal lives of individuals.

3. **Mass hysteria** is a form of dispersed collective behavior that occurs when a large number of people react with strong emotions and self-destructive behavior to a real or perceived threat; many sociologists believe this behavior is best described as a panic with a dispersed audience.

4. Fads and Fashions.
   a. A fad is a temporary but widely copied activity enthusiastically followed by large numbers of people.
   b. Fashion is a currently valued style of behavior, thinking or appearance. Fashion also applies to art, music, drama, literature, architecture, interior design, and automobiles, among other things.

5. **Public opinion** consists of the attitudes and beliefs communicated by ordinary citizens to decision makers (as measured through polls and surveys based on interviews and questionnaires).
   a. Even on a single topic, public opinion will vary widely based on characteristics such as race, ethnicity, religion, region of the country, urban or rural residence, social class, education level, gender, and age.
   b. As the masses attempt to influence elites and visa versa, a two-way process occurs with the dissemination of **propaganda** -- information provided by individuals or groups that have a vested interest in furthering their own cause or damaging an opposing one.

II. SOCIAL MOVEMENTS

A. A **social movement** is an organized group that acts consciously to promote or resist change through collective action.

B. Types of Social Movements
   1. Reform movements seek to improve society by changing some specific aspect of the social structure.
   2. Revolutionary movements seek to bring about a total change in society. **Terrorism** is the calculated unlawful use of physical force or threats of violence against persons or property in order to intimidate or coerce a government, organization, or individual for the purpose of gaining some political, religious, economic or social objective.
   3. Religious movements seek to produce radical change in individuals and typically are based on spiritual or supernatural belief systems.
   4. Alternative movements seek limited change in some aspect of people's behavior (e.g., a movement that attempts to get people to abstain from drinking alcoholic beverages).
   5. Resistance movements seek to prevent or to undo change that already has occurred.
C. Causes of Social Movements
   1. Relative deprivation theory asserts that people who suffer relative deprivation are likely to feel that a change is necessary and to join a social movement in order to bring about that change.
   2. According to **Neal Smelser**'s value-added theory, six conditions are necessary and sufficient to produce social movements when they combine or interact in a particular situation:
      a. Structural conduciveness
      b. Structural strain
      c. Spread of a generalized belief
      d. Precipitating factors
      e. Mobilization for action
      f. Social control factors
   3. Resource mobilization theory focuses on the ability of a social movement to acquire resources (money, time and skills, access to the media, etc.) and mobilize people to advance the cause.
   4. Emerging Perspectives.
      a. Emerging perspectives based on resource mobilization theory emphasize ideology and legitimacy of movements as well as material resources.
      b. Recent theories based on an interactionist perspective focus on the importance of the symbolic presentation of a problem both to participants and the general public.
      c. Alan Scott notes that, over the past two decades, "new social movements" have placed more emphasis on quality of life issues than earlier movements that focused primarily on economic issues.

        d.   Examples of already existing "new social movements" include ecofeminism and environmental justice movements.

            (1)   According to ecofeminists, patriarchy is a root cause of environmental problems because it contributes to a belief that nature is to be possessed and dominated, rather than treated as a partner.

            (2)   Environmental justice movements focus on the issue of environmental racism -- the belief that a disproportionate number of hazardous facilities (including industries such as waste disposal/treatment and chemical plants) are placed in low-income areas populated primarily by people of color.

D.   Stages in Social Movements

    1.   In the preliminary stage, widespread unrest is present as people begin to become aware of a threatening problem. Leaders emerge to agitate others into taking action.

    2.   In the coalescence stage, people begin to organize and start making the threat known to the public. Some movements become formally organized at local and regional levels.

    3.   In the institutionalization stage, an organizational structure develops, and a paid staff (rather than volunteers) begins to lead the group.

III.   SOCIAL CHANGE: MOVING INTO THE TWENTY-FIRST CENTURY

A.   The Physical Environment and Change: Changes in the physical environment often produce changes in the lives of people; in turn, people can make dramatic changes in the physical environment, over which we have only limited control.

B.   Population and Change: Changes in population size, distribution, and composition affect the culture and social structure of a society and change the relationships among nations.

C.   Technology and Change: Advances in communication, transportation, science, and medicine have made significant changes in people's lives, especially in developed nations; however, these changes also have created the potential for new disasters, ranging from global warfare to localized technological disasters at toxic waste sites.

D.   Social Institutions and Change: During the twentieth century, many changes have occurred in the family, religion, education, the economy, and the political system that will follow us into the twenty-first century.

E.   Changes in physical environment, population, technology, and social institutions operate together in a complex relationship, sometimes producing consequences we must examine by using our sociological imagination.

ANALYZING AND UNDERSTANDING THE BOXES

After reading the chapter and studying the outline, re-read the four boxes and write down key points and possible questions for class discussion.

Sociology and Everyday Life -- "How Much Do You Know About Collective Behavior and Environmental Issues?"

Key Points:

Discussion Questions:

1.

2.

3.

Sociology and Media -- "Getting Attention for a Cause"

Key Points:

Discussion Questions:

1.

2.

3.

Sociology in Global Perspective -- "Environmental Hazards as a Global Concern"

Key Points:

Discussion Questions:

1.

2.

3.

Sociology and Law -- "Water Rights"

Key Points:

Discussion Questions:

1.

2.

3.

PRACTICE TEST

MULTIPLE CHOICE QUESTIONS

Select the response that best answers the question or completes the statement:

1.   All of the following are factors that contribute to the likelihood that collective behavior will occur, <u>except</u>: (p. 619)
      a.   the presence of deviant behavior.
      b.   structural factors that increase the chances of people responding in a particular way.
      c.   timing.
      d.   a breakdown in social control mechanisms and a corresponding feeling of normlessness.

2.   According to sociologist William A. Gamson, the millions of people who agreed with a ban on nuclear weapons but did not go to a rally in New York City are: (p. 620)
      a.   political "moochers."
      b.   free riders.
      c.   social parasites.
      d.   uncommitted activists.

3.   Casual crowds are: (p. 620)
      a.   comprised of people who specifically come together for a scheduled event and thus share a common focus.
      b.   situations that provide an opportunity for the expression of some strong emotion.
      c.   comprised of people who happen to be in the same place at the same time.
      d.   comprised of people who are so intensely focused on a specific purpose or object that they may erupt into violent or destructive behavior.

4. Post-game celebrations that turn violent are an example of: (p. 623)
   a. protest crowds.
   b. riots.
   c. conventional crowds.
   d. panics.

5. When the residents of Love Canal picketed to halt the work of the state, which they considered to be insufficient to solve the problem of toxic chemical waste, they were engaging in a: (p. 623)
   a. protest crowd.
   b. riot.
   c. conventional crowd.
   d. panic.

6. Convergence theory focuses on: (p. 625)
   a. the social-psychological aspects of collective behavior, including how moods, attitudes, and behavior are communicated.
   b. how social unrest is transmitted by a process of circular reaction.
   c. the importance of social norms in shaping crowd behavior.
   d. the shared emotions, goals, and beliefs many people bring to crowd behavior.

7. All of the following statements regarding the emergent norm theory are true, except: (p. 626)
   a. Emergent norm theory is based on the interactionist perspective.
   b. Sociologists using the emergent norm approach seek to determine how individuals in a given collectivity develop an ·understanding of what is going on, how they construe these activities, and what types of norms are involved.
   c. Emergent norm theory points out that crowds sometimes are irrational.
   d. Emergent norm theory originated with sociologists Ralph Turner and Lewis Killian.

8. Rumors, gossip, fashions, and fads are examples of _____ behavior. (p. 627)
   a. mob
   b. mass
   c. irrational
   d. casual

9. A form of dispersed collective behavior that occurs when a large number of people react with strong emotions and self-destructive behavior to a real or perceived threat is known as: (p. 628)
   a. mob behavior.
   b. mass behavior.
   c. mass hysteria.
   d. contagious behavior.

10. "Streaking" -- students taking off their clothes and running naked in public -- in the 1970s is an example of a: (p. 628)
    a. panic.
    b. trend.
    c. fashion.
    d. fad.

11. According to the text, _____ is information provided by individuals or groups that have a vested interest in furthering their own cause or damaging an opposing one. (p. 630)
    a. propaganda
    b. public opinion
    c. political rhetoric
    d. a press release

12. Which of the following statements regarding social movements is true? (p. 630)
    a. Social movements are more likely to develop in preindustrial societies where there is an acceptance of traditional beliefs and practices.
    b. Social movements have become institutionalized and are a part of the political mainstream.
    c. Social movements make democracy less accessible to excluded groups.
    d. Social movements offer "outsiders" an opportunity to have their voices heard.

13. All of the following are types of social movements, except: _____ movements. (p. 631)
    a. alternative
    b. dissident
    c. religious
    d. revolutionary

14. According to relative deprivation theory: (p. 634)
    a. people who are satisfied with their present condition are more likely to seek social change.
    b. certain conditions are necessary for the development of a social movement.
    c. people who feel that they have been deprived of their "fair share" are more likely to feel that change is necessary and to join a social movement.
    d. some people bring more resources to a social movement than others.

15. _____ theory is based on the assumption that six conditions, including structural conduciveness and structural strain, must be present for the development of a social movement. (p. 635)
    a. Value-added
    b. Relative deprivation
    c. Resource mobilization
    d. Emergent norm

16. _____ theory focuses on the ability of members of a social movement to acquire resources and mobilize people in order to advance their cause. (p. 635)
   a. Value-added
   b. Relative deprivation
   c. Resource mobilization
   d. Emergent norm

17. According to sociologist Alan Scott, "new social movements" have focused on: (p. 638)
   a. the availability of money, people's time and skills, and access to the media.
   b. the level of relative deprivation experienced by movement participants.
   c. the extent to which structural strain is present in a society or community.
   d. quality-of-life issues such as the environment.

18. In the _____ stage of a social movement, people begin to organize and to publicize the problem. (p. 639)
   a. preliminary
   b. coalescence
   c. institutionalization
   d. deinstitutionalization

19. The belief that a disproportionate number of hazardous facilities are placed in low-income areas populated by people of color is known as: (p. 639)
   a. environmental racism.
   b. environmental justice.
   c. reverse environmentalism.
   d. racial pollution.

20. All of the following statements regarding natural disasters are true, except: (p. 641)
   a. Major natural disasters can dramatically change the lives of people.
   b. Trauma that people experience from disasters may outweigh the actual loss of physical property.
   c. Natural disasters are not affected by human decisions.
   d. Disasters may become divisive elements that tear communities apart.

TRUE-FALSE QUESTIONS

T F 1. Collective behavior lacks an official division of labor, hierarchy of authority, and established rules and procedures. (pp. 618-619)

T F 2. People are more likely to act as a collectivity when they believe it is the only way to fight those with greater power and resources. (p. 620)

T   F   3.   People gathered for religious services and graduation ceremonies are examples of casual crowds. (p. 620)

T   F   4.   Through the efforts of Love Canal residents, all of the toxic waste has been removed from their community. (p. 622)

T   F   5.   Protest crowds engage in activities intended to achieve specific political goals. (p. 623)

T   F   6.   Sociologist Robert E. Park was the first U.S. sociologist to investigate crowd behavior. (p. 625)

T   F   7.   According to emergent norm theory, people with similar attributes find a collectivity of like-minded persons with whom they can express their underlying personal tendencies. (p. 625)

T   F   8.   For mass behavior to occur, people must be in close proximity geographically. (p. 626)

T   F   9.   Public opinion does not always translate into action by decision makers in government and industry or by individuals. (p. 630)

T   F   10.   Grassroots environmental movements are an example of reform movements. (p. 631)

T   F   11.   The 1995 bombing of the Federal Building in Oklahoma City was the worst incident of domestic terrorism in U.S. history. (p. 633)

T   F   12.   Revolutionary movements also are referred to as expressive movements. (p. 633)

T   F   13.   Movements based on relative deprivation are most likely to occur when people have unfulfilled rising expectations. (p. 635)

T   F   14.   Gains made by social movements may not be long lasting. (p. 639)

T   F   15.   One of the major concerns in the twenty-first century is likely to be the availability of water. (p. 640)

SOCIOLOGY IN OUR TIMES: DIVERSITY ISSUES

1.   Have you ever participated in a social movement? If so, what types of issues were involved in the movement? Were the issues related to race/ethnicity, class, gender, age, sexual orientation, religion, or disability? Why are so many social movements intertwined with one or more of these characteristics?

347

2. Watch television news programs for examples of collective behavior or social movements. Who are the participants? What are the major issues involved? When did the behavior occur? Where did it occur? Why did participants engage in this behavior? Does the behavior of participants have any impact on your life? Why or why not?

3. Can you cite evidence that environmental racism exists in your community or city? Why do some sociologists believe that environmental justice is a pressing social issue?

4. Based on your race/ethnicity, gender, class, and age, how do you think you will be affected by changes in the physical environment, population, technology, and social institutions as we approach the twenty-first century?

CHAPTER SEVENTEEN CROSSWORD PUZZLE

For those who enjoy crossword puzzles, here is a puzzle that contains words and names from Chapter Seventeen. Working the puzzle will help you in reviewing the chapter. The answers appear on page 352.

ACROSS

1. [9 down] _____: an organized group that acts consciously to promote or resist change through collective action
4. _____ opinion: the attitudes and beliefs communicated by ordinary citizens to decision makers
9. See 1 across
10. An unsubstantiated report on an issue or subject
11. [8 down] _____ argued that people in the middle and lower classes follow fashion because it is *fashion*
12. Box 17.2 demonstrates how social movements use the ____ to gain attention to their causes
13. Sociologist Robert _____ asserted that social unrest is transmitted by a process of circular reaction
14. He argued that people are more likely to engage in antisocial behavior in [19 down] because they are anonymous and feel invulnerable
17. Initials of 13 across
18. A form of crowd behavior that occurs when a large number of people react to a real or perceived threat with strong emotions and self-destructive behavior
23. He asserted that fashion served mainly to institutionalize conspicuous consumption among the wealthy
24. See 1 down
25. See 13 across
27. Tabloid newspapers ___ magazines such as the *National Enquirer* and *People* are sources of much contemporary gossip
28. _____ behavior: voluntary, often spontaneous activity that is engaged in by a large number of people and typically violates dominant group norms and values

DOWN

1. _____ [24 across]: collective behavior that takes place when people (who often are geographically separated from one another) respond to the same event in much the same way
2. Sociologist Kai ____ describes a "new species of trouble" – environmental problems that contaminate, pollute, befoul, and taint, rather than just creating wreckage
3. He and Lewis Killian assert that crowds develop their own definition of a situation and establish norms for behavior that fits the occasion
5. Information provided by individuals or groups that have a vested interest in furthering their own cause or damaging an opposing one
6. He suggested that, when a change in the material culture occurs in a society, a period of cultural lag follows
7. He suggested a classic "trickle down" theory to describe the process by which members of the lower classes emulate the fashions of the upper class
8. See 11 across
14. Crowd: a relatively ____ number of people who are in one another's immediate vicinity
15. Violent crowd behavior that is fueled by deep-seated emotions but not directed at one specific target
16. A highly emotional crowd whose members engage in, or are ready to engage in, violence against a specific target
18. A casual crowd plays no ___ in the action
19. See 14 across
20. "Free riders" are people who enjoy the benefits produced by some group ____ though they have not helped support it
21. Movements based on _____ism often use tactics such as bombings, kidnappings, hostage taking, etc.
22. Clark Mc_____ and Ronald T. Wohlstein added protest crowds to the types of crowds identified by [11 across]
23. Ms. Capek and Mr. Clayman have these initials in common

349

ANSWERS TO THE PRACTICE TEST, CHAPTER 17

Answers to Multiple Choice Questions

1. a   All of the following are factors that contribute to the likelihood that collective behavior will occur, <u>except</u>: the presence of deviant behavior. (p. 619)

2. b   According to sociologist William A. Gamson, the millions of people who agreed with a ban on nuclear weapons but did not go to a rally in New York City are free riders. (p. 620)

3. c   Casual crowds are comprised of people who happen to be in the same place at the same time. (p. 620)

4. b   Post-game celebrations that turn violent are an example of riots. (p. 623)

5. a   When the residents of Love Canal picketed to halt the work of the state, which they considered to be insufficient to solve the problem of toxic chemical waste, they were engaging in a protest crowd. (p. 623)

6. d   Convergence theory focuses on the shared emotions, goals, and beliefs many people bring to crowd behavior. (p. 625)

7. c   All of the following statements regarding the emergent norm theory are true, <u>except</u>: emergent norm theory points out that crowds sometimes are irrational. (p. 626)

8. b   Rumors, gossip, fashions, and fads are examples of mass behavior. (p. 627)

9. c   A form of dispersed collective behavior that occurs when a large number of people react with strong emotions and self-destructive behavior to a real or perceived threat is known as mass hysteria. (p. 628)

10. d   "Streaking" -- students taking off their clothes and running naked in public -- in the 1970s is an example of a fad. (p. 628)

11. a   According to the text, propaganda is information provided by individuals or groups that have a vested interest in furthering their own cause or damaging an opposing one. (p. 630)

12. d   Which of the following statements regarding social movements is true? Social movements offer "outsiders" an opportunity to have their voices heard. (p. 630)

13. b   All of the following are types of social movements, <u>except</u>: dissident movements. (p. 631)

14. c   According to relative deprivation theory, people who feel that they have been deprived of their "fair share" are more likely to feel that change is necessary and to join a social movement. (p. 634)

15. a   Value-added theory is based on the assumption that six conditions, including structural conduciveness and structural strain, must be present for the development of a social movement. (p. 635)

16. c   Resource mobilization theory focuses on the ability of members of a social movement to acquire resources and mobilize people in order to advance their cause. (p. 635)

17. d According to sociologist Alan Scott, "new social movements" have focused on quality-of-life issues such as the environment. (p. 638)
18. b In the coalescence stage of a social movement, people begin to organize and to publicize the problem. (p. 639)
19. a The belief that a disproportionate number of hazardous facilities is placed in low-income areas populated by people of color is known as environmental racism. (p. 639)
20. c All of the following statements regarding natural disasters are true, <u>except</u>: natural disasters are not affected by human decisions. (p. 641)

Answer to True-False Questions

1. True (pp. 618-619)
2. True (p. 620)
3. False -- People gathered for religious services and graduation ceremonies are examples of conventional crowds. (p. 620)
4. False -- Although the efforts of Love Canal residents did make people more aware of environmental pollution, more than 21,000 tons of toxic chemical waste and the remains of 239 contaminated homes still remain underneath a 40-acre grassy landfill in Niagara, New York, that once again is "home" to 160 recent buyers of renovated houses in the area. (p. 622)
5. True (p. 623)
6. True (p. 625)
7. False -- According to convergence theory, people with similar attributes find a collectivity of like-minded persons with whom they can express their underlying personal tendencies. (p. 625)
8. False -- Mass behavior often takes place when people who are geographically separated from one another respond to the same event in much the same way (for example, rumor, fashion, or fad).
9. True (p. 630)
10. True (p. 631)
11. True (p. 633)
12. False -- Religious movements also are referred to as expressive movements. (p. 633)
13. True (p. 635)
14. True (p. 639)
15. True (p. 640)

```
M O V E M E N T   O P I N I O N   S   H
A       R       U   R       G       I   E
A S O C I A L   R U M O R       B L U M E R
S       K       N   P       U       M   B
        S       M E D I A   P A R K   E   E
L E B O N       R   I       N       L   R
A       N       R   G       M           T
R E P           I   P A N I C   E   O   T
G       P       O   A   D   R   V E B L E N
E       H Y S T E R I A   O   E       R
        A           T   W   N   P A R K
S       I           A N D           O
C O L L E C T I V E     S M E L S E R
```